T0140341

Studies in Computational Intelligence

Volume 1079

Series Editor

Janusz Kacprzyk, Polish Academy of Sciences, Warsaw, Poland

The series "Studies in Computational Intelligence" (SCI) publishes new developments and advances in the various areas of computational intelligence—quickly and with a high quality. The intent is to cover the theory, applications, and design methods of computational intelligence, as embedded in the fields of engineering, computer science, physics and life sciences, as well as the methodologies behind them. The series contains monographs, lecture notes and edited volumes in computational intelligence spanning the areas of neural networks, connectionist systems, genetic algorithms, evolutionary computation, artificial intelligence, cellular automata, self-organizing systems, soft computing, fuzzy systems, and hybrid intelligent systems. Of particular value to both the contributors and the readership are the short publication timeframe and the world-wide distribution, which enable both wide and rapid dissemination of research output.

Indexed by SCOPUS, DBLP, WTI Frankfurt eG, zbMATH, SCImago.

All books published in the series are submitted for consideration in Web of Science.

Hang Chen · Zhengjun Liu

Editors

Recent Advanced in Image Security Technologies

Intelligent Image, Signal, and Video Processing

 Springer

Editors
Hang Chen
School of Space Information
Space Engineering University
Beijing, China

Zhengjun Liu
School of Physics
Harbin Institute of Technology
Harbin, China

ISSN 1860-949X   ISSN 1860-9503 (electronic)
Studies in Computational Intelligence
ISBN 978-3-031-22811-7   ISBN 978-3-031-22809-4 (eBook)
https://doi.org/10.1007/978-3-031-22809-4

This Springer imprint is published by the registered company Springer Nature Switzerland AG
The registered company address is: Gewerbestrasse 11, 6330 Cham, Switzerland

Preface

With the rapid development of modern photography technology, the application of real-time imaging transmission has been widely used both in commercial and military area, such as resource monitoring, remote surgery, military target detection. Meanwhile, the image security technologies are becoming more and more important in transmission and storage process. The commonly used video and images can be protected with different kinds of security technology, which help the owners to prevent unauthorized distribution and use. Furthermore, the image (video) security technologies have remarkable military significance and application in the fields of space remote sensing, navigation, and space command operation.

Over the past two decades, the information security technologies have attracted significant attention and led to extensive study of individuals, companies, institutes, and government sectors all around the world. Image encryption technique is one way to convert the secret information to an unreadable format or a confused image which is hard to understand. However, different from the text message, the image data has some special characteristics like high capacity, redundancy, and high correlation among pixels. Particularly, the optical information security technology has become increasingly attractive since the double random phase encoding (DRPE) is proposed by Refregier and Javidi in 1995 due to its high-speed parallel processing. The key motivation of using optics and photonics in information security is that the various complex degrees of freedom. Besides, some other encryption technologies are also used in image cryptosystem, such chaotic/hyperchaotic maps, pixels encoding technology, FPGA-based implementation, embedded hardware implementation, which can extremely improve the security of the encryption schemes. In the past decade, the artificial intelligence algorithms are also employed in the image cryptosystem, such as CNN and SVM.

This book provides the readers with a comprehensive overview of principles, methodologies, and recent advances in image and video security processing (contains but not limited to encryption, hiding, and watermarking) using latest technologies through a collection of high-quality chapters. These chapters present original works of famous researches from several countries. This book is a valuable reference for scientists, engineers, scientific researchers, postgraduate students, and senior under-graduate students with majors in image, signal, and video processing. To enlighten the reader, we would like to divide this book into eight chapters:

The first chapter -"Optical Information Cryptosystems Based on Structured Phase Encoding."

The second chapter -"Encryption/Decryption with Optical Transform."

The third chapter -"Optical Cryptosystems Based on Spiral Phase Modulation."

The fourth chapter -"Image Cryptosystem for Different Kinds of Image by Using Improved Arnold Map."

The fifth chapter -"Image Encryption Using a Chaotic/Hyperchaotic Multidimensional Discrete System."

The sixth chapter -"Video Cryptosystem Using Chaotic Systems."

The seventh chapter -"Neural Network Image Restoration Techniques."

Beijing, China   Hang Chen
Harbin, China   Zhengjun Liu

Acknowledgements The editors would like to take this opportunity to express their sincere grat-itude to the authors of the chapters for extending their wholehearted support in sharing some of their latest results. Without their significant contribution, this book volume could not have fulfilled its mission. The editor is very grateful to Editor-in-Chief Prof. Thomas Ditzinger for his help on our book. All the authors are indebted to the editorial staff at Springer, for their hard working and selfish attitude on helping our book plan.

Contents

Optical Information Cryptosystems Based on Structured Phase Encoding

Muhammad Rafiq Abuturab

Abstract The structured phase mask (SPM) is a diffractive optical element. It mainly comprises a radial Hilbert phase mask (RHPM), a Fresnel zone plate phase mask (FZPM), a spiral zone plate phase mask (SZPM), and a chaotic spiral phase mask (CSPM). It is easier to align with setup axis and it is itself a multiple-key device. Optical cryptosystems based on RHPM and FZPM in the gyrator transform domain have been presented. In these cryptosystems, the construction parameters are used as encryption keys. An asymmetric information cryptosystem based on SZPM and optical interference principle has been studied. Finally, asymmetric multiple-information cryptosystem based on CSPM and random spectrum decomposition has been discussed. In asymmetric cryptosystems, the construction parameters are used as decryption keys. Hybrid optoelectronic systems can be used for the implementation of cryptosystems. Numerical simulations demonstrate the efficiency and viability of the optical methods.

1 Introduction

Information security is of paramount importance in communication systems, where the rate of information dissemination is growing phenomenally. The security of traditional methods is improved using stronger cryptographic algorithms. High-speed computers can reduce the time to decrypt an encrypted message. When the length of the encrypting key becomes longer, the processing speed becomes slower. So, optical information security techniques have been widely applied in the field of information security, owing to their multiple parameters and parallel processing capability [1]. The security of a cryptosystem depends on the length of the security key/mask and its randomness. Thus, the security key plays a significant role in achieving a higher level of security of a cryptosystem. In optical cryptosystems, the keys are mostly phase-only functions called random phase masks (RPMs). The RPM is positioned in

M. R. Abuturab (✉)
Department of Physics, Muzaffarpur Institute of Technology, Muzaffarpur 842003, India
e-mail: rafiq.abuturab@gmail.com

© The Author(s), under exclusive license to Springer Nature Switzerland AG 2023
H. Chen and Z. Liu (eds.), *Recent Advanced in Image Security Technologies*,
Studies in Computational Intelligence 1079,
https://doi.org/10.1007/978-3-031-22809-4_1

the Fourier plane of a classical Fourier processor to encrypt the Fourier spectrum of an image/data [2]. It has two important practical disadvantages. Speckle noise and the sensitivity of the alignment in the decoding step. First, speckle masks produce higher spatial frequency components, which cause interference patterns to change rapidly in certain local areas in the holographic plane. Second, a misalignment of RPM in the decryption process causes to fail in retrieving the original image/data. Therefore, precise alignment of an optical cryptosystem based on a $4f$ correlator is one of the important practical issue.

The structured phase masks were introduced to overcome the misalignment problem in the decryption stage. The radially symmetric Hilbert transform (RHT) was introduced and applied to the two-dimensional edge enhanced images of arbitrarily shaped input objects [3]. It is easier to align with the setup axis [4]. The toroidal zone plate (TZP) was reported and analyzed its sensitivity to misalignment [5]. The Fresnel zone plate (FZP) was proposed in gyrator transform domain to solve alignment problems in optical cryptosystem based on a $4f$ correlator [6, 7]. The spiral zone plate (SZP) was put forward [8]. The optical cryptosystem based on optical interference principle and spiral zone plate phase encoding was proposed to attain precise alignment [9]. The chaotic structured phase mask (CSPM) was proposed in gyrator transform domain to solve axis alignment [10, 11].

In this chapter, optical cryptosystems based on structured phase masks have been reported. This chapter starts with a brief introduction to structured phase masks. The optical techniques for image/data security systems based on structured phase masks, algorithms, and optical implementations have been discussed. Finally, numerical simulation results have been made to test the security, validity, and efficiency of the optical techniques.

The rest of the paper is organized as follows. In Sect. 1, the structured phase mask and its types are described. In Sect. 2, optical gyrator transform is presented. In Sect. 3, the optical information cryptosystems based on structured phase masks are discussed. The chapter ends with concluding remarks.

2   Structured Phase Mask

The structured phase mask (SPM) is an alternative to the random phase mask (RPM). The SPM mainly includes a radial Hilbert phase mask (RHPM), Fresnel zone plate phase mask (FZPM), spiral zone plate phase mask (SZPM), and chaotic spiral phase mask (CSPM). The SPM is defined by a shape given by a specific configuration. It offers remarkable advantages. First, it is very difficult to replicate as it is a diffractive optical element. Second, it is easier to position during the decoding process as it provides its own centering mark (on-axis focal ring when illuminated by a parallel beam). Third, its diffraction efficiency can be manipulated depending on the requirements of the optical system. Finally, it is itself a multiple-key device as it contains multiple keys in a single-phase mask.

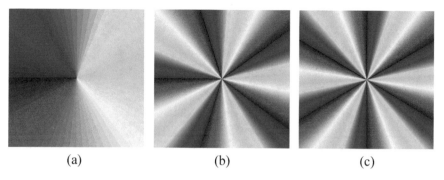

$$(a) \qquad\qquad (b) \qquad\qquad (c)$$

Fig. 1 Radial Hilbert mask for **a** $m = 1$, **b** $m = 5$, and **c** $P = 6$

2.1 Radial Hilbert Phase Mask

The Hilbert transform is useful for image-processing applications because it forms an image that is edge enhanced relative to an input object. However, 1D Hilbert transform enhances edges along only a single direction. Mathematically, the 1D Hilbert transform is the convolution of the object with $-1/\pi x$ function and this function is the Fourier transform of a signum function in x-direction. The 2D Hilbert masks can be produced by forming the product of two Hilbert masks $H_m(u)$ and $H_m(v)$. However, these masks retain basic x, y symmetry. The 2D Hilbert masks can also be produced to avoid the basic x, y symmetry by making a radial mask in which opposite halves of any radial line have a relative phase difference of $m\pi$ rad. Therefore, for each radial line, we have the equivalent of a one-dimensional Hilbert transform of an order m. The RHPM is given by

$$H_m(r, \phi) = \exp(im\phi) \tag{1}$$

where the variables (r, ϕ) denote polar coordinates on the plane of the spatial light modulator. m is the order of RHT, which is used as a security key. Figure 1 shows three numerically generated RHPMs constructed for three different orders.

2.2 Fresnel Zone Plate Phase Mask

The complex field amplitude distribution of FZPM is given by

$$Z(r, \phi) = \exp\left(\frac{-i\pi r^2}{\lambda f}\right) \tag{2}$$

where r is the radius and f is the focal length of the FZPM. λ is the illuminating wavelength. The polar coordinates (r, ϕ) are related to the Cartesian coordinates by

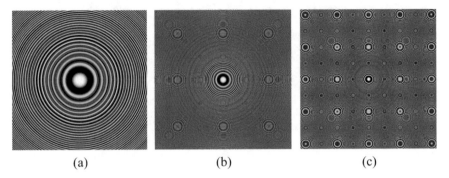

Fig. 2 Fresnel zone plate phase mask. a $f_1 = 3$ cm, $r_1 = 0.1$ mm, $\lambda_{r_1} = 635.0$ nm, $\lambda_{g_1} = 531.0$ nm, and $\lambda_{b_1} = 473.0$ nm; **b** $f_2 = 4$ cm, $r_2 = 0.2$ mm, $\lambda_{r_2} = 650.0$ nm, $\lambda_{g_2} = 545.0$ nm, and $\lambda_{b_2} = 450.0$ nm; **c** $f_3 = 5$ cm, $r_3 = 0.3$ mm, $\lambda_{r_3} = 632.8$ nm, $\lambda_{g_3} = 532.0$ nm, and $\lambda_{b_3} = 488.0$ nm

$r^2 = x^2 + y^2$, $\phi = \tan^{-1}(y/x)$. The optical axis is assumed to be coincident with the propagation axis. The parameters of FZPM (λ, f, and r) are used as security keys. Figure 2 shows three numerically generated FZPMs constructed for three different focal lengths, radii, red, green, and blue wavelengths (λ_r, λ_g, λ_b).

2.3 Spiral Zone Plate Phase Mask

The SZPM is produced by multiplying the radial Hilbert phase function $H_m(r, \phi)$ with the Fresnel zone plate phase function $Z(r, \phi)$.

$$S(r, \phi) = H_m(r, \phi)Z(r, \phi) = \exp\left[i\left(m\phi - \frac{\pi}{\lambda f}r^2\right)\right]. \tag{3}$$

The focusing ring of the mask can be aligned with the setup axis. The parameters of SPM (m, λ, f, and r) are used as security keys. Figure 3 shows three numerically generated FZPMs constructed for three different orders, focal lengths, radii, red, green, and blue wavelengths (λ_r, λ_g, λ_b).

2.4 Chaotic Spiral Zone Plate Phase Mask

The one-dimensional (1D) non-linear chaos function is a logistic map, and its iterative form is expressed as

$$x_{n+1} = px_n(1 - x_n) \tag{4}$$

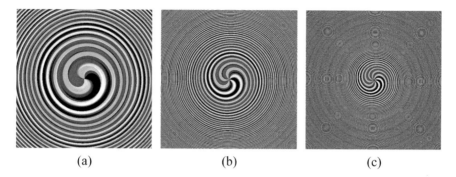

$$(a) \qquad\qquad (b) \qquad\qquad (c)$$

Fig. 3 Spiral zone plate phase mask. a $m_{R_1} = 1, m_{G_1} = 2, m_{B_1} = 3, f_1 = 1.5$ cm, $r_1 = 1$ mm, $\lambda_{r_1} = 635.0$ nm, $\lambda_{g_1} = 531.0$ nm and $\lambda_{b_1} = 473.0$ nm; **b** $m_{R_2} = 4, m_{G_2} = 5, m_{B_2} = 6, f_2 = 2$ cm, $r_2 = 1.5$ mm, $\lambda_{r_2} = 650.0$ nm, $\lambda_{g_2} = 545.0$ nm and $\lambda_{b_2} = 450.0$ nm; **c** $m_{R_3} = 7, m_{G_3} = 8$, $m_{B_3} = 9, f_3 = 2.5$ cm, $r_3 = 2$ mm, $\lambda_{r_3} = 632.8$ nm, $\lambda_{g_3} = 532.0$ nm and $\lambda_{b_3} = 488.0$ nm

where p is called bifurcation parameter: $0 < p < 4$. $x_n \in [0, 1]$ and x_0 are iterative and initial values.

Let the chaotic random phase mask (CRPM) be of size $M \times N$ pixels. Initially, the 1D random value sequence is produced by using Eq. (4) as $X = \{x_1, x_2, ..., x_{(M \times N)+k}\}$, $x_i \in (0, 1)$ and k is any chosen integer. Subsequently, the 2D matrix is generated by rearranging the sequence X as $Y = \{y_{i,j} | i = 1, 2, ..., M + k; j = 1, 2, ..., N + k\}$, $y_{i,j} \in (0, 1)$. CRPM is defined as [10, 11]

$$C(x, y) = \exp[i2\pi y_{i,j}(x, y)]. \tag{5}$$

The CSPM is produced by multiplying the spiral phase function $S(r, \phi)$ with the chaotic random phase function $C(x, y)$

$$M(x, y) = C(x, y)S(r, \phi) = \exp\left\{i\left[2\pi y_{i,j}(x, y) + m\theta - \frac{\pi}{\lambda f}r^2\right]\right\}. \tag{6}$$

The parameters of CSPM (x_0, p, k, m, λ, f, and r) are used as security keys. Figure 4 shows three numerically generated CSPMs constructed for three different values of parameters of CSPM.

3 Gyrator Transform

The optical gyrator transform (GT) operation G^α of a two-dimensional function $f_i(x_i, y_i)$ with parameter α, known as rotation angle, is defined as [12]

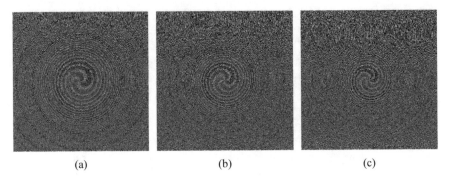

Fig. 4 Chaotic Spiral phase mask. **a** $x_{01} = 0.35$, $p_1 = 3.960$, $k_1 = 2000$, $m_{R_1} = 2$, $m_{G_1} = 3$, $m_{B_1} = 4$, $\lambda_{R_1} = 632.8$ nm, $\lambda_{G_1} = 532$ nm, $\lambda_{B_1} = 488$ nm, $f_1 = 0.030$ m, and $r_1 = 0.0020$ m; **b** $x_{02} = 0.40$, $p_2 = 3.970$, $k_2 = 2100$, $m_{R_2} = 3$, $m_{G_2} = 4$, $m_{B_2} = 5$, $\lambda_{R_2} = 635$ nm, $\lambda_{G_2} = 531$ nm, $\lambda_{B_2} = 473$ nm, $f_2 = 0.035$ m, and $r_2 = 0.0025$ m; **c** $x_{03} = 0.45$, $p_3 = 3.980$, $k_3 = 2200$, $m_{R_3} = 4$, $m_{G_3} = 5$, $m_{B_3} = 6$, $\lambda_{R_3} = 650$ nm, $\lambda_{G_3} = 545$ nm, $\lambda_{B_3} = 450$ nm, $f_3 = 0.040$ m, and $r_3 = 0.0030$ m

$$f_o(x_o, y_o) = G^{\alpha}[f_i(x_i, y_i)](x_o, y_o) = \iint f_i(x_i, y_i) K_{\alpha}(x_i, y_i; x_o, y_o) dx_i dy_i \quad (7)$$

where (x_i, y_i) and (x_o, y_o) are the input and output plane coordinates, respectively.

The Kernel of GT is given as

$$K_{\alpha}(x_i, y_i; x_o, y_o) = \frac{1}{|\sin \alpha|} \exp\left(i2\pi \frac{(x_o y_o + x_i y_i) \cos \alpha - (x_i y_o + x_o y_i)}{\sin \alpha} \right). \quad (8)$$

The possible values of α are in the following interval $0 \leq \alpha < 2\pi$. For $\alpha = 0$ and $\alpha = \pi$, it corresponds to the identity transform and reverse transform, respectively. For, $\alpha = \pi/2$ and $\alpha = 3\pi/2$, it reduces to the direct and inverse Fourier transform with the rotation of the coordinates at $\pi/2$. The inverse GT corresponds to the GT at the rotation angle $-\alpha$. For other angles α, the Kernel of GT $K_{\alpha}(x_i, y_i; x_o, y_o)$ has a constant amplitude and a hyperbolic phase structure. The GT is additive and periodic with respect to parameter α: $G^{\alpha} G^{\beta} = G^{\alpha+\beta}$ and $G^{\alpha+2\pi} = G^{\alpha}$. As G^{α} and $G^{2\pi-\alpha}$ are reciprocal transforms, it is used in two-dimensional image processing.

The experimental setup of optical GT comprises three generalized lenses (denoted as L_1, L_2 and L_3) and two fixed equal intervals z [13]. Each generalized lens is an assembled set of two identical plano-convex cylindrical lenses of the same power. The first and third identical generalized lenses are rotated with respect to each other. The focal distance f_1 equals the distance z between two consecutive generalized lenses of the setup. The second generalized lens of focal length $f_2(= z/2)$ is fixed. The third generalized lens compensates the undesirable phase modulation introduced by rotation of the first and third generalized lenses. The variation of transformation angles α is achieved by proper rotation of these lenses. GT operation for different transformation angles can be performed by only proper rotation of cylindrical lenses, where the distance between them and input–output planes are fixed. Therefore, GT optical systems do not require axial movements.

4 Optical Information Cryptosystem Using Structured Phase Encoding

The optical color image/data encryption based on structured phase mask has been discussed. First, the color image/data is encrypted by using the RHPM in GT domain RHPM and GT do not require axial movements [4]. Second, the color image/data is encrypted by using the FZPM in GT domain. FZPM and GT avoid problems relating to misalignment [6, 7]. Third, the color image/data is encrypted by using the SZPM and optical interference principle. Its optical setup eases stringent alignment of spiral phase-only masks (SOPMs) [9]. Finally, the color image/data is encrypted by using the CSPM and random spectrum decomposition in which fewer optical components are employed in the decryption optical setup [11].

4.1 Optical Information Cryptosystem Using Radial Hilbert Phase Mask in Gyrator Transform Domain

In this section, a color image security system based on discrete cosine transform (DCT) and RHPM in GT domain is introduced. In the encryption technique, each channel is encrypted independently by using first RPM and DCT at the image plane, and then performed the first GT. The transformed image is again encrypted by using second RPM and DCT at GT plane, transmitted through RHPM, and then executed second GT. The system parameters of RHPM and GT in each channel are used as encryption keys to enlarge the key space. RHPM and GT in the proposed method do not require axial movements. Numerical simulations have been presented to verify the efficiency and viability of the proposed cryptosystem.

DCT is employed in the image encryption because of two advantages. First, DCT changes the spatial distribution of pixel value of an image. Second, it defines real number field. Thus, output data can be encoded with only real numbers.

The original color image $f(x, y)$ is split into R, G, and B channels denoted as $f_R(x_i, y_i)$, $f_G(x_i, y_i)$ and $f_B(x_i, y_i)$, respectively. For simplicity, $f_c(x_i, y_i)$ is considered as a color channel, where $c = R, G, B$.

In encryption process, $f_c(x_i, y_i)$ is multiplied by the first RPM $P_{1c}(x_i, y_i)$ encoded by the first DCT, and then performed the first GT at rotation angle α_c. The transformed image is multiplied by second RPM $P_{2c}(x, y)$, encoded by second DCT, modulated by RHPM $R_c(x, y)[= H_m(r, \phi)]$, and then executed the second GT at rotation angle β_c. The encrypted image is obtained as

$$E_c(x_o, y_o) = G_c^\beta \left\{ C_c \left\{ G_c^\alpha [C_c[f_c(x_i, y_i) P_{1c}(x_i, y_i)]] P_{2c}(x, y) \right\} R_c(x, y) \right\}. \quad (9)$$

Decryption process is inverse of the encryption process. In the decryption process, the encrypted image $E_c(x_o, y_o)$ is inverse gyrator transformed at rotation angle β_c, multiplied by the conjugate of RHPM $R_c^*(x, y)$, and then performed inverse DCT. The resultant image multiplied by the conjugate of second RPM $P_{2c}^*(x, y)$ is inverse gyrator transformed at rotation angle α_c, executed inverse DCT, and then multiplied by the conjugate of first RPM $P_{1c}^*(x, y)$. The decrypted image is retrieved as

$$D_c(x_i, y_i) = C_c^{-1} \left\{ G_c^{-\alpha} \left\{ C_c^{-1} \left[G_c^{-\beta} [E_c(x_o, y_o)] R_c^*(x, y) \right] P_{2c}^*(x, y) \right\} \right\} P_{1c}^*(x_i, y_i). \quad (10)$$

The optoelectronic hybrid setup of the proposed encryption process is shown in Fig. 5. The dotted block compring lenses L_1, L_2 and L_3 denotes the first optical GT and that compring lenses L_1', L_2' and L_3' represents the second optical GT. For convenience, only R channel is mentioned. The spatial distribution of pixel value of R channel is changed digitally by the first DCT, attached to a first RPM, displayed on the first Spatial Light Modulator SLM_1 at input plane, and then optically transformed by first GT. The spatial distribution of pixel value of the transformed image is changed digitally by the second DCT, attached to the second RPM, displayed on the second Spatial Light Modulator SLM_2 at GT plane, transmitted through RHT, and then optically transformed by second GT. The resultant image is superimposed on the plane reference beam to produce a holographic interference fringe, which is captured and recorded as an off-axis hologram by charged couple device (CCD) camera, and digitally processed by a computer system. Each color channel independently recorded and processed by the same method are combined to form an encrypted color image. The decrypted image is obtained by reverse of the encryption process.

Numerical simulations have been performed on a Matlab 9.10 (R2021a) platform. A color image having $512 \times 512 \times 3$ pixels and 24 bits is regarded as an original image as shown in Fig. 6a. The RHPM of order $P = 5$ is shown in Fig. 1b. The rotation

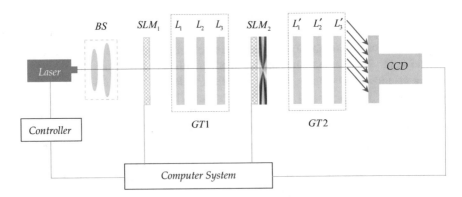

Fig. 5 Optoelectronic hybrid encryption architecture

angles of the first GT and second GT are, respectively, α_c ($\alpha_R = 3°$, $\alpha_G = 4°$, $\alpha_G = 5°$) and β_c ($\beta_R = 6°$, $\beta_G = 7°$, and $\beta_G = 8°$). The encrypted image is shown in Fig. 6b. The decrypted image with all the correct keys is shown in Fig. 6c. The retrieved images without second RPM, with $\Delta\alpha_c = 0.5°$, and with $\Delta\beta_c = 0.1°$ are, respectively, shown in Fig. 6d–f. The results show that the difference between input and decrypted images cannot be discerned visually. Thus, the security system is robust to second RPM and sensitive to small variations in rotation angles of GT.

To evaluate the quality of decrypted image, the correlation coefficient (CC) values between original R, G, and B channels and their corresponding retrieved channels calculated against variation of m, α, and β are, respectively, plotted in Fig. 7a–c. For brevity, $\alpha_c = \beta_c = 5°$.

In all cases, when the variation of the system parameter approximates to correct value, the related CC value becomes one, when the parameter departs from the correct value slightly, the CC value decreases rapidly. Noticeably, the cryptosystem is very sensitive to tiny variation of the system parameters.

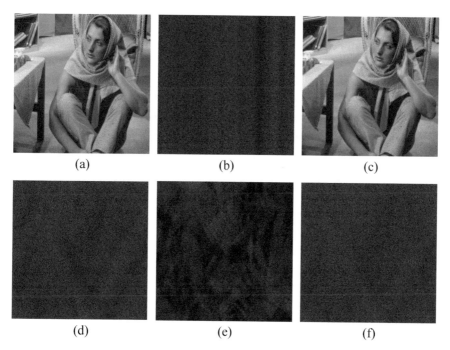

Fig. 6 **a** Original image, **b** encrypted image, **c** decrypted image with all the correct keys, **d** decrypted image without second RPM, **e** decrypted image with $\Delta\alpha_c = 0.5°$, **f** decrypted image with $\Delta\beta_c = 0.1°$

Fig. 7 CC as a function of deviation of **a** m, **b** α, **c** β

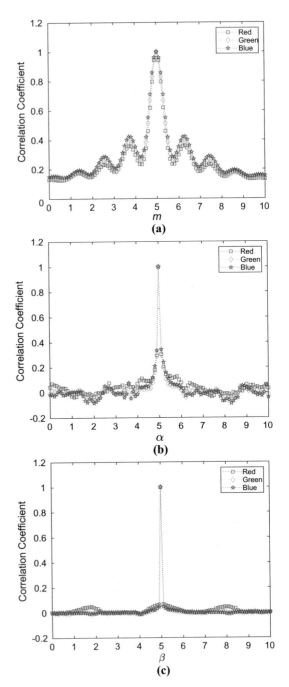

4.2 Optical Information Cryptosystem Using Fresnel Zone Plate Phase Mask in Gyrator Transform Domain

In this section, a color image security system based on DCT and FZPM in GT domain is introduced. In the encryption process, the input color image is decomposed into R, G, and B channels. Each channel is individually encrypted by changing its spatial distribution of pixel value by the first DCT, modulated by first FZPM, and then performed the first GT. The transformed image is again encrypted by second DCT, modulated by second FZPM, and then executed the second GT. Decryption process is inverse of encryption process. The construction parameters of FZPM and angle parameters of GT in each channel are principal encryption keys. The proposed architecture does not require axial movements. Numerical simulations have been conducted to confirm the security and viability of the proposed method.

The original color image $f(x, y)$ is decomposed into R, G, and B channels denoted as $f_R(x_i, y_i)$, $f_G(x_i, y_i)$ and $f_B(x_i, y_i)$, respectively. For brevity, $f_c(x_i, y_i)$ is considered as a color channel, where $c = R, G, B$.

In encryption process, $f_c(x_i, y_i)$ is encoded by the first DCT, modulated by first FZPM $F_{c1}(x_i, y_i)$, and then performed the first GT at rotation angle α_c. The transformed image is encoded by the second DCT, modulated by second FZPM $F_{c2}(x, y)$, and then executed second GT at rotation angle β_c.

The encrypted image is obtained as

$$E_c(x_o, y_o) = G_c^\beta \big\{ \{ C_c \big[G_c^\alpha [C_c[f_c(x_i, y_i)]F_{c1}(x_i, y_i)] \big] \} F_{c2}(x, y) \big\} \qquad (11)$$

where $F_{c1}(x_i, y_i) = \exp\left(\frac{-i\pi r_{c1}^2}{\lambda_{c1} f_{c1}}\right)$ and $F_{c2}(x, y) = \exp\left(\frac{-i\pi r_{c2}^2}{\lambda_{c2} f_{c2}}\right)$

Decryption process is inverse of the encryption process. In the decryption process, the encrypted image $E_c(x_o, y_o)$ is inverse gyrator transformed at rotation angle β_c, multiplied by the conjugate of second FZPM $F_{2c}^*(x, y)$ and then performed inverse DCT. The resultant image is inverse gyrator transformed at rotation angle α_c, multiplied by the conjugate of second FZPM $F_{1c}^*(x, y)$ and then executed inverse DCT. The decrypted image is retrieved as

$$D_c(x_i, y_i) = C_c^{-1} \big\{ G_c^{-\alpha} \{ C_c^{-1} \big[G_c^{-\beta} [E_c(x_o, y_o)] F_{c2}^*(x, y) \big] \} F_{c2}^*(x_i, y_i) \big\} \qquad (12)$$

The optoelectronic hybrid setup of the proposed encryption process is shown in Fig. 8. The dotted block compring lenses L_1, L_2, and L_3 denotes the first optical GT and that compring lenses L_1', L_2', and L_3' represents the second optical GT. For the sake of clarity, only R channel is described. The spatial distribution of pixel value of the red channel is changed digitally by the first DCT and displayed on the SLM_1 at input plane. The obtained distribution is transmitted through first FZPM, and then transformed optically by first GT. Now the spatial distribution of pixel value of the resulting image is changed digitally by the second DCT and displayed on the SLM_2 at GT plane. The complex distribution is transmitted through second FZPM,

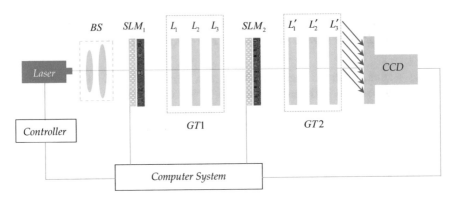

Fig. 8 Optoelectronic hybrid encryption system

and then transformed optically by second GT. The resultant image is superimposed on the plane reference beam to produce a holographic interference fringe, which is captured and recorded as an off-axis hologram by charged couple device (*CCD*) camera, and digitally processed by a computer system. R, G, and B channels are independently recorded and processed by the same technique are combined to form an encrypted color image. The reverse of the encryption procedure gives a decrypted image.

Numerical simulations have been performed on a Matlab 9.10 (R2021a) environment. The original color image with $512 \times 512 \times 3$ pixels and 24 bits is shown in Fig. 9a. The first FZPM of focal length $f_1 = 4$ cm, radius $r_1 = 0.2$ mm, red wavelength $\lambda_{r_1} = 650.0$ nm, green wavelength $\lambda_{g_1} = 545.0$ nm, and blue wavelength $\lambda_{b_1} = 450.0$ nm is shown in Fig. 2b. The second FZPM of focal length $f_2 = 5$ cm, radius $r_2 = 0.3$ mm, red wavelength $\lambda_{r_2} = 632.8$ nm, green wavelength $\lambda_{g_2} = 532.0$ nm, and blue wavelength $\lambda_{b_2} = 488.0$ nm is shown in Fig. 2c. The rotation angles of the first GT and second GT are, respectively, α_c ($\alpha_R = 3°$, $\alpha_G = 4°$, $\alpha_G = 5°$) and β_c ($\beta_R = 6°,$, $\beta_G = 7°$, and $\beta_G = 8°$). The encrypted and decrypted images with all correct keys are, respectively, shown in Fig. 9b, c. The decrypted image with deviation in system parameters have been studied. The decrypted images with $\Delta f_2 = 1 \times 10^{-3}$ m, $\Delta r_2 = 1 \times 10^{-5}$ m and $\Delta \lambda_c = 5 \times 10^{-9}$ m are, respectively, shown in Fig. 9d–f. The decrypted image without First FZPM is shown in Fig. 9g. The decrypted image with $\Delta \alpha_c = 0.01°$ and $\Delta \beta_c = 0.01°$ are, respectively, shown in Fig. 9h, i. The results show that any original information cannot be recognized visually. Thus, the security system is robust to first FZPM. The system is sensitive to small variations in parameters of second FZPM and rotation angles of GT.

To evaluate the quality of decrypted images, the correlation coefficient (CC) values between original R, G, and B channels and their corresponding recovered channels calculated versus deviation of f_2, r_2, λ_2, α, and β, are, respectively, plotted in Fig.10a–e. For simplicity, $\alpha_c = \beta_c = 5°$.

In all cases, when the deviation of the system parameter approaches to exact value, the corresponding CC value reaches one, while the parameter departs from the exact

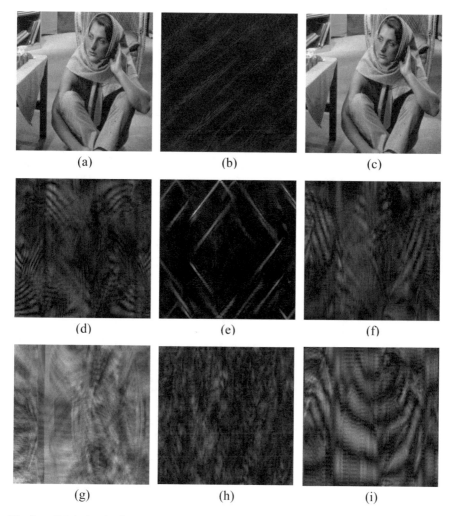

Fig. 9 **a** Original color image, **b** encrypted image, **c** decrypted image with all correct keys, **d** decrypted image with $\Delta f_2 = 1 \times 10^{-3}$ m, **e** decrypted image with $\Delta r_2 = 1 \times 10^{-5}$ m, **f** decrypted image with $\Delta \lambda_{c2} = 5 \times 10^{-9}$ m, **g** decrypted image without First FZPM, **h** decrypted image with $\Delta \alpha_c = 0.01°$, **i** decrypted image with $\Delta \beta_c = 0.01°$

value marginally, the CC value decreases quickly. Clearly, the security system is very sensitive to minute deviation in the system parameters.

Fig. 10 CC as a function of deviation of **a** f_2, **b** r_2, **c** λ_2, **d** α, and **e** β

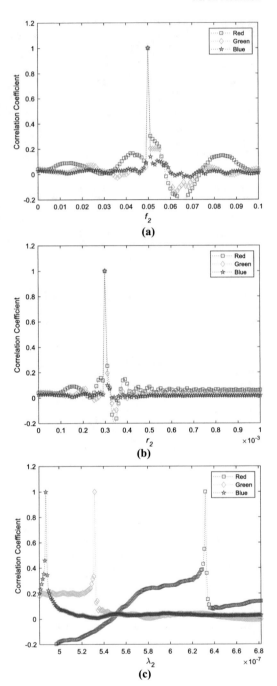

Fig. 10 (continued)

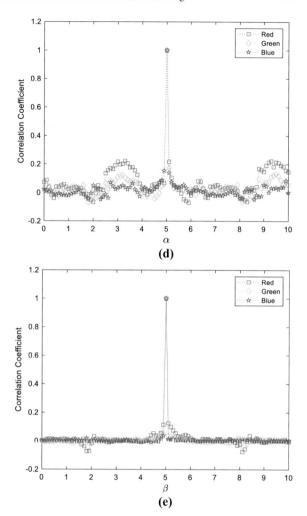

(d)

(e)

4.3 Asymmetric Information Cryptosystem Based on Spiral Zone Plate Phase Mask and Optical Interference Principle

In this section, an asymmetric color image cryptosystem using the optical interference principle and spiral phase encoding (SPM) is introduced. In the encryption scheme, the color image is dissociated into R, G, and B channels. Each channel is encrypted into an SPM and analytically produced two spiral phase-only masks (SPOMs). The two SPOMs are then phase-truncated to get two encrypted images and amplitude-truncated to generate two asymmetric phase keys. The four construction parameters of the SPM (or third SPOM) are the order, the wavelength, the focal length, and the radius. The proposed method has advantages over earlier methods. First, similar to the earlier proposal, this technique also employs three POMs as SPOMs for encryption and decryption procedures. However, the SPM as the third SPOM provides four additional keys. Second, two asymmetric keys generated during the encryption process are also used as extra keys. So, all the additional keys are produced without using an iterative algorithm or post-processing of SPOMs to remove the silhouette problem. The two SPOMs, the two asymmetric keys, and four keys in the third SPOM can be assigned to eight different authorized users. So, an unauthorized user cannot recover the original image by using a combination of any seven keys. The optical system avoids strict alignment of the POMs during the decryption process. Numerical simulations have been carried out to validate robustness and feasibility of the proposed technique.

In this method, an original color image $f(x, y)$ is segregated into R, G, and B channels denoted as $f_R(x_i, y_i)$, $f_G(x_i, y_i)$ and $f_B(x_i, y_i)$, respectively. For briefness, $f_c(x_i, y_i)$ is considered as a color channel, where $c = R, G, B$.

The normalized input image $f_c(x_i, y_i)$ is multiplied by an SPM $S_c(r, \theta)$ and a complex function $I_c(x_i, y_i)$ is obtained as [14]

$$I_c(x_i, y_i) = \sqrt{f_c(x_i, y_i)} S_c(r, \theta) \tag{13}$$

Let $S_{c_1}(x, y)$, $S_{c_2}(x, y)$ and $S_{c_3}(x, y)$ be three SPOMs. Then the relationship of the three SPOMs with the complex function $I_c(x_i, y_i)$ is expressed as

$$I_c(x_i, y_i) = F\left[S_{c_1}(x, y) + S_{c_2}(x, y) + S_{c_3}(x, y)\right] \tag{14}$$

where $F[]$ denotes the Fourier transform operator.

If $E_c(x, y) = F^{-1}[I_c(x_i, y_i)]$ denotes the interference pattern distribution. Then its relationship with the three SPOMs is

$$S_{c_1}(x, y) + S_{c_2}(x, y) + S_{c_3}(x, y) = F^{-1}[I_c(x_i, y_i)] = E_c(x, y) \tag{15}$$

where $F^{-1}[]$ symbolizes the inverse Fourier transform operator.

The three SPOMs are defined as

$$S_{c_1}(x, y) = \exp\left[i s_{c_1}(x, y)\right] \tag{16}$$

$$S_{c_2}(x, y) = \exp\left[i s_{c_2}(x, y)\right] \tag{17}$$

$$S_{c_3}(x, y) = S_c(r, \theta) \tag{18}$$

If $E'_c(x, y) = E_c(x, y) - S_{c_3}(x, y)$ represents the interference pattern distribution of $S_{c_1}(x, y)$ and $S_{c_2}(x, y)$, where the phase information of $E_c(x, y)$ is not preserved in $E'_c(x, y)$ [15]. That is, two SPOMs $S_{c_1}(x, y)$ and $S_{c_2}(x, y)$ are analytically obtained from $E'_c(x, y)$ whereas the third SPOM $S_{c_3}(x, y)$ is analytically generated by Eq. (3).

According to the nature of SPOMs,

$$\left[E_c'(x, y) - S_{c_1}(x, y)\right]\left[E_c'(x, y) - S_{c_1}(x, y)\right]^* = 1 \tag{19}$$

where the subscript * indicates the complex conjugate. Therefore, the phase distributions of the $S_{c_1}(x, y)$ and $S_{c_2}(x, y)$ are given by

$$s_{c_1}(x, y) = \arg\left[E_c'(x, y)\right] - \arccos\left[\left|E_c'(x, y)/2\right|\right] \tag{20}$$

$$s_{c_2}(x, y) = \arg\left[E_c'(x, y)\right] + \arccos\left[\left|E_c'(x, y)/2\right|\right] \tag{21}$$

where the operators arg[] and | | are, respectively, the phase and the modulus of the function.

The phase truncations of both SPOMs $S_{c_1}(x, y)$ and $S_{c_2}(x, y)$ produce two asymmetric ciphertexts [16].

$$E_{c_1}(x, y) = PT\left[S_{c_1}(x, y)\right] \tag{22}$$

$$E_{c_2}(x, y) = PT\left[S_{c_2}(x, y)\right] \tag{23}$$

where $PT[]$ represents the operator of the phase truncation.

Similarly, the amplitudes truncations of both SPOMs $S_{c_1}(x, y)$ and $S_{c_2}(x, y)$ generate first and second asymmetric phase keys.

$$k_{c_1}(x, y) = AT\left[S_{c_1}(x, y)\right] \tag{24}$$

$$k_{c_2}(x, y) = AT\left[S_{c_2}(x, y)\right] \tag{25}$$

where $AT[]$ indicates the operator of amplitude truncation.

The decryption process is much simpler compared to the encryption process.

$$D_{c_1}(x, y) = E_{c_1}(x, y)k_{c_1}(x, y) \tag{26}$$

$$D_{c_2}(x, y) = E_{c_2}(x, y)k_{c_2}(x, y) \tag{27}$$

$$f_c(x_i, y_i) = PT\left\{F\left[D_{c_1}(x, y) + D_{c_2}(x, y) + S_{c_3}(x, y)\right]\right\} \tag{28}$$

In the proposed optical interference-based security system, the encryption process is implemented digitally while the decryption process can be realized optically. The optoelectronic design for the decryption system is shown in Fig. 11. The stringent alignment of two SPOMs in two different arms during the experiment is circumvented by using a single SLM. The encoded $E_{c_1}(x, y)$ and $E_{c_2}(x, y)$ are, respectively, multiplied by $k_{c_1}(x, y)$ and $k_{c_2}(x, y)$. Then the summation of recovered two SPOMs, and SPM (as the third SPOM) is displayed on the SLM. The laser beam is expanded with a beam expander to illuminate SLM. The resultant phase distribution obtained from the summation of three SPOMs is optically Fourier transformed and then recorded by a CCD camera in the interference plane. The process is repeated to retrieve R, G, and B channels, which are combined by the computer system to get the original color image.

Numerical simulations have been carried out on a Matlab 9.10 (R2021a) platform to test the feasibility and security of the cryptosystem. The original color image having size $512 \times 512 \times 3$ pixels and 24 bits is shown in Fig. 12a. The SPM as SPOM3 generated by using its construction parameters of the orders ($m_R = 1, m_G = 2, m_B = 3$), wavelengths ($\lambda_R = 635$ nm, $\lambda_G = 531$ nm, $\lambda_B = 473$ nm), focal length $f = 1.5$ cm, and radius $r = 0.1$ mm, is displayed in Figs. 3a or 12d, where subscripts R, G, and B denote red, green, and blue color channels, respectively. The phase distributions of $S_{c_1}(x, y)$ and $S_{c_2}(x, y)$ are, respectively, illustrated in Fig. 12b, c. The encrypted image is shown in Fig. 12e. The decrypted image with all correct keys is displayed in Fig. 12f.

The decrypted images without $S_{c_1}(x, y)$, without $S_{c_2}(x, y)$, without $k_{c_1}(x, y)$ and without $k_{c_2}(x, y)$ but with all correct keys are depicted in Fig. 13a−d, respectively. The decrypted results demonstrate that the original image cannot be recovered even if only one key is unknown to an unauthorized user. If an error is introduced in one parameter of $S_{c_3}(x, y)$ but with all the other correct keys then the sensitivity of error in, $\Delta m = 0.05$, $\Delta \lambda = 0.3$ nm, $\Delta f = 0.0005$ cm, and $\Delta r = 0.00002$ mm, is very high since no valuable information can be obtained from the corresponding decrypted images as shown in Fig. 14a−d. These results confirm that the construction parameters of $S_{c_3}(x, y)$ are extremely sensitive and they alone cause great difficulty in duplicating the decryption keys. Consequently, the proposed system provides a higher degree of security.

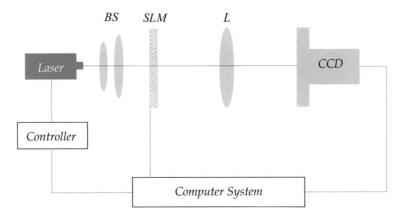

Fig. 11 Optoelectronic hybrid decryption system

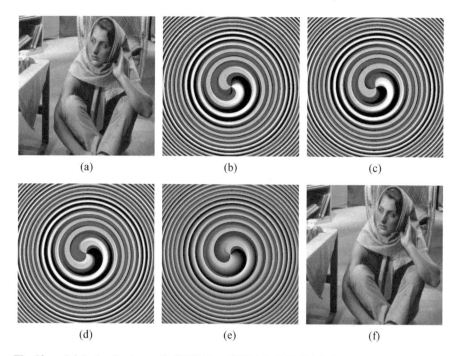

Fig. 12 **a** Original color image, **b** SPOM 1, **c** SPOM 2, **d** SPOM 3, **f** encrypted image, and **g** decrypted image with all correct keys

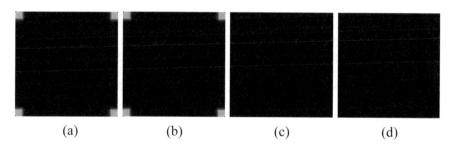

Fig. 13 Recovered images **a** without SPOM 1, **b** without SPOM 2, **c** without first asymmetric phase key, **d** without second asymmetric phase key

To evaluate the quality of decrypted images, the correlation coefficient (CC) curve of R, G, and B channels versus variation of m, λ, f, and r of $S_{c_3}(x, y)$ are, respectively, plotted in Fig.15a–d, while other keys are correct.

It can be seen in each curve diagram, the CC is one when the key is correct and declines steeply when the key slightly departs from the correct value owing to the extreme sensitivity of the parameters of SPM (as SPOM 3).

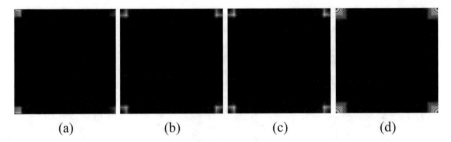

(a)   (b)   (c)   (d)

Fig. 14 Decryption results of SPM (as SPOM 3) with error in **a** $\Delta m_c = 0.05$, **b** $\Delta \lambda_c = 0.3$ nm, **c** $\Delta f = 0.0005$ cm, **d** $\Delta r = 0.00002$ mm

4.4 Asymmetric Multiple Information Cryptosystem Based on Chaotic Spiral Phase Mask and Random Spectrum Decomposition

In this section, a new asymmetric multiple-image encryption (MIE) based on chaotic spiral phase mask (CSPM) and random spectrum decomposition is introduced. In the encryption system, each channel of a secret color image is first multiplied with a CSPM and then a gyrator transform is implemented. The gyrator spectrum is split into two complex-valued masks randomly. A similar process is used to multiple secret images to get their corresponding first and second complex-valued masks. Finally, the first and second masks of each channel are separately added to generate the first and second complex ciphertexts, respectively. Simulation results have been performed to verify the effctiveness and feasibility of the proposed MIE scheme.

The proposed method has advantages over reported methods. Compared with Refs. [17, 18], more sensitive and robust parameters of CSPM $(x_0, p, k, m, \lambda, f,$ and $r)$ are used as decryption keys instead of encryption keys and hence enhance the security levels. Compared with Refs. [10, 19, 20], the different secret images are encrypted by different CSPMs using different parameters as decryption keys that cannot be identified by unauthorized users. Moreover, fewer optical components are employed in the decryption optical setup. Compared with Ref. [10], a time-consuming iterative algorithm is not involved in the proposed algorithm.

Suppose $f_1(x_i, y_i)$ denotes the intensity distribution of the first secret color image of the first authorized user to be encrypted. The first color image is decomposed into $R, G,$ and B channels indicated as $f_{R_1}(x_i, y_i)$, $f_{G_1}(x_i, y_i)$ and $f_{B_1}(x_i, y_i)$, respectively. For brevity, the only red channel is exemplified. The function $I_{R_1}(x, y)$ can be constructed as

$$I_{R_1}(x, y) = G^{\alpha}\left[\sqrt{f_{R_1}(x_i, y_i)} . S_{R_1}(x_i, y_i)\right] \tag{29}$$

where $S_{R_1}(x_i, y_i)$ is CSPM of the red channel of the first color image.

A complex number can be regarded as a position vector in a two-dimensional Cartesian coordinate system. The real part (Re) and the imaginary part (Im) are

Fig. 15 CC curve for deviation of **a** m, **b** λ, **c** f, and **d** r

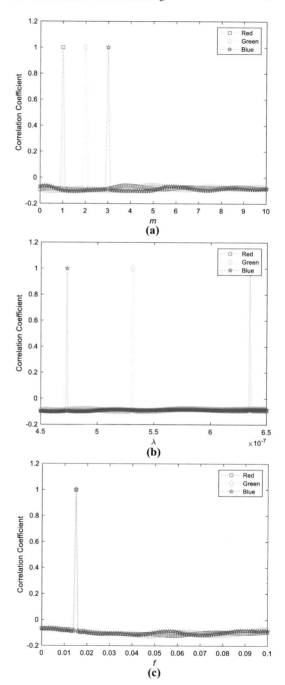

Fig. 15 (continued)

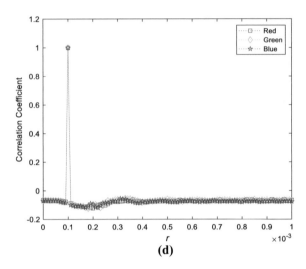

(d)

represented by horizontal and vertical components, respectively. As shown in Fig. 16, the function $I_{R_1}(x, y)$ is randomly decomposed into two complex-valued masks $P_{R_{11}}(x, y)$ and $P_{R_{12}}(x, y)$ [18]. $\alpha_{R_1}(x, y)$ and $\beta_{R_1}(x, y)$ are given by

$$\alpha_{R_1}(x, y) = \beta_{R_1}(x, y) = \left(2\pi y_{i,j}(x, y) + m_{R_1}\theta - \frac{\pi}{\lambda_{R_1} f} r^2\right) \qquad (30)$$

The amplitude and phase of Eq. (29) are written as

Fig. 16 Basic vector operation of random decomposition

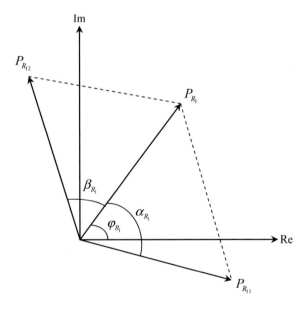

$$A_{R_1}(x, y) = \left| I_{R_1}(x, y) \right| \tag{31}$$

and

$$\phi_{R_1}(x, y) = \arg\left[I_{R_1}(x, y) \right] \tag{32}$$

The operators $|\ |$ and $\arg[\ ]$ denote the modulus and phase, respectively.

Using simple geometrical deduction, complex-valued masks $P_{R_{11}}(x, y)$ and $P_{R_{12}}(x, y)$ are defined as

$$P_{R_{11}}(x, y) = \frac{A_{R_1}(x, y).\sin\left[\beta_{R_1}(x, y)\right]}{\sin\left[\alpha_{R_1}(x, y) + \beta_{R_1}(x, y)\right]} \cdot \exp\left\{ -i\left[\alpha_{R_1}(x, y) - \phi_{R_1}(x, y)\right] \right\} \tag{33}$$

$$P_{R_{12}}(x, y) = \frac{A_{R_1}(x, y).\sin\left[\alpha_{R_1}(x, y)\right]}{\sin\left[\alpha_{R_1}(x, y) + \beta_{R_1}(x, y)\right]} \cdot \exp\left\{ i\left[\beta_{R_1}(x, y) + \phi_{R_1}(x, y)\right] \right\} \tag{34}$$

The identical procedure is applied to the rest of $n - 1$ red channel of secret images $f_{R_2}(x_i, y_i), f_{R_3}(x_i, y_i), \ldots, f_{R_n}(x_i, y_i)$ of second, third, ..., $n - 1$ authorized users to obtain corresponding complex-valued masks, $\left[P_{R_{21}}(x, y) \text{ and } P_{R_{22}}(x, y)\right], \left[P_{R_{31}}(x, y) \text{ and } P_{R_{32}}(x, y)\right], \ldots, \left[P_{R_{n1}}(x, y) \text{ and } P_{R_{n2}}(x, y)\right]$. Finally, $P_{R_{11}}(x, y)$, $P_{R_{21}}(x, y)$, $P_{R_{31}}(x, y)$, ..., and $P_{R_{n1}}(x, y)$ are added to get the first complex ciphertext $P_R^1(x, y)$. $P_{R_{12}}(x, y), P_{R_{22}}(x, y), P_{R_{32}}(x, y), \ldots,$ and $P_{R_{n2}}(x, y)$ are added to obtain second complex ciphertext $P_R^2(x, y)$.

The complex ciphertexts $P_R^1(x, y)$ and $P_R^2(x, y)$ are expressed as

$$P_R^1(x, y) = P_{R_{11}}(x, y) + P_{R_{21}}(x, y) + P_{R_{31}}(x, y) + \cdots + P_{R_{n1}}(x, y) \tag{35}$$

$$P_R^2(x, y) = P_{R_{12}}(x, y) + P_{R_{22}}(x, y) + P_{R_{32}}(x, y) + \cdots + P_{R_{n2}}(x, y) \tag{36}$$

The individual decryption keys $K_{R_{11}}(x, y)$ and $K_{R_{12}}(x, y)$ of the first authorized user are obtained as

$$K_{R_{11}}(x, y) = -\left[P_{R_{21}}(x, y) + P_{R_{31}}(x, y) + \cdots + P_{R_{n1}}(x, y)\right] \tag{37}$$

$$K_{R_{12}}(x, y) = -\left[P_{R_{22}}(x, y) + P_{R_{32}}(x, y) + \cdots + P_{R_{n2}}(x, y)\right] \tag{38}$$

The sum of $P_R^1(x, y)$, $K_{R_{11}}(x, y)$, $P_R^2(x, y)$, and $K_{R_{12}}(x, y)$ are inverse gyrator transformed to retrieve original image as

$$f_{R_1}(x_i, y_i) = \left| G^{-\alpha}\left\{ \left[P_R^1(x, y) + K_{R_{11}}(x, y)\right] + \left[P_R^2(x, y) + K_{R_{12}}(x, y)\right]\right\}\right| \tag{39}$$

This process is repeated for $f_{G_1}(x_i, y_i)$ and $f_{B_1}(x_i, y_i)$ to the first secret image. Similarly, $n - 1$ secret images can be decrypted.

The proposed encryption process is performed digitally, whereas the decryption process can be executed digitally or optically. The optoelectronic setup for the optical decryption process for R channel is shown in Fig. 17. The first encrypted red-channel $P_R^1(x, y)$ and first decryption key $K_{R_{n1}}(x, y)$ are added, displayed on the two closely assembled first spatial light modulators SLM_1 (for amplitude modulation) and SLM_1' (for phase modulation), and then illuminated with a collimated laser beam 1. The second encrypted red-channel $P_R^2(x, y)$ and second decryption key $K_{R_{n2}}(x, y)$, are added, displayed on the two closely assembled second spatial light modulators SLM_2 (for amplitude modulation) and SLM_2' (for phase modulation) and then illuminated with a collimated laser beam 2. The two spectra interfere with each other at the gyrator plane. The intensity of the retrieved red channel is recorded by a CCD camera and stored in the computer system. The retrieved blue and green channels separately recorded and stored by the same method are combined to produce an original color image.

Numerical simulations have been executed on a Matlab 9.10 (2021a) environment to calculate the effectiveness and security of the proposed method. The three color

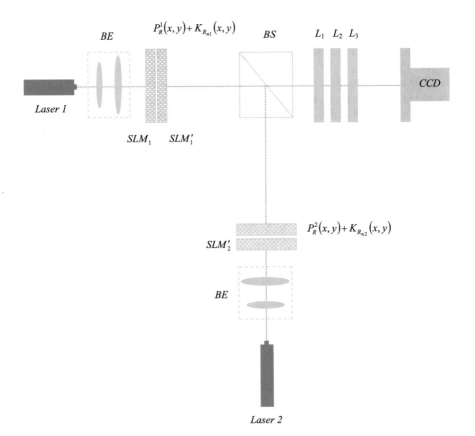

Fig. 17 Optoelectronic hybrid decryption setup

images of each of the size $512 \times 512 \times 3$ pixels to be hidden are Barbara, Father, and Ali assigned to authorized users 1, 2, and 3, as shown in Fig. 18a–c, respectively. The parameters of CSPF for first, second, and third users are ($x_{01} = 0.35$, $p_1 = 3.960$, $k_1 = 2000$, $m_{R_1} = 2$, $m_{G_1} = 3$, $m_{B_1} = 4$, $\lambda_{R_1} = 632.8$ nm, $\lambda_{G_1} = 532$ nm, $\lambda_{B_1} = 488$ nm, $f_1 = 0.030$ m, $r_1 = 0.0020$ m), ($x_{02} = 0.40$, $p_2 = 3.970$, $k_2 = 2100$, $m_{R_2} = 3$, $m_{G_2} = 4$, $m_{B_2} = 5$, $\lambda_{R_2} = 635$ nm, $\lambda_{G_2} = 531$ nm, $\lambda_{B_2} = 473$ nm $f_2 = 0.035$ m, $r_2 = 0.0025$ m), and ($x_{03} = 0.45$, $p_3 = 3.980$, $k_3 = 2200$, $m_{R_3} = 4$, $m_{G_3} = 5$, $m_{B_3} = 6$, $\lambda_{R_3} = 650$ nm, $\lambda_{G_3} = 545$ nm, $\lambda_{B_3} = 450$ nm $f_3 = 0.040$ m, $r_3 = 0.0030$ m), respectively. The angles of the GT for first, second, and third users are $\alpha_1 = 0.40°$ $\alpha_2 = 0.50°$ and $\alpha_3 = 0.60°$), respectively. Figures 4a–4c display first, second, and third CRPMs for first, second, and third users, respectively. Figures 18d, and 18e depict the first ciphertext $P_R^1(x, y)$, and second ciphertext $P_R^2(x, y)$, respectively. The reconstructed first, second, and third images with all correct keys are, respectively, demonstrated in Fig. 18f–h.

Figure 19a–f illustrate first key of first authorized user, second key of first authorized user, first key of second authorized user, second key of second authorized user, first key of third authorized user, and second key of third authorized user, respectively. The decrypted images without first individual key of first image, without second individual key of first image, without first individual key of second image, without second individual key of second image, without first individual key of third image, and without second individual key of third image are, respectively, shown in Fig. 19g–l. It can be seen that if any of the correct security keys is not available to unauthorized users, the information of the secret images can not be recovered and hence demonstrates the robustness of the proposed technique against brute force attacks.

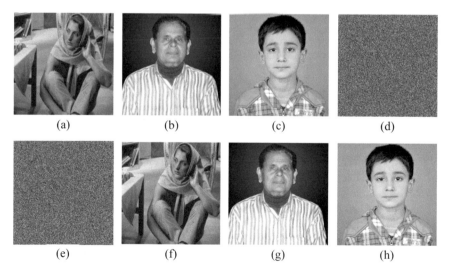

Fig. 18 Secret color images: **a** Barbara, **b** father, **c** Ali; phase of **d** first ciphertext, **e** second ciphertext; retrieved image with correct keys: **f** Barbara, **g** father, **h** Ali

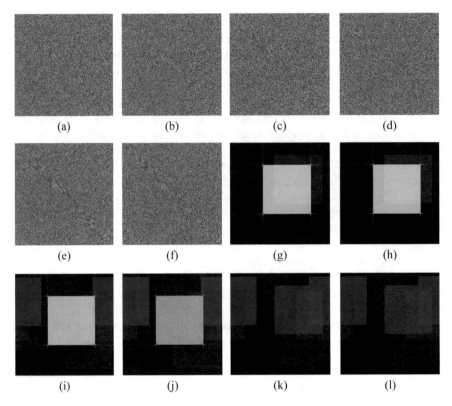

Fig. 19 Phase of individual decryption keys: **a** first key of first authorized user, **b** second key of first user, **c** first key of second authorized user, **d** second key of second authorized user, **e** first key of third authorized user, and **f** second key of third authorized user; decrypted image **g** without first individual key of first image, **h** without second individual key of first image, **i** without first individual key of second image, **j** without second individual key of second image, **k** without first individual key of third image, **l** without second individual key of third image

The sensitivity of system parameters of the first secret image of authorized user I has been examined. The recovered secret images with deviation in parameters of $\Delta x_0 = 1 \times 10^{-16}$, $\Delta k = 1$, $\Delta p = 1 \times 10^{-15}$, $\Delta m = 0.01$, $\Delta \lambda = 0.1$ nm, $\Delta f = 0.5 \times 10^{-4}$ m, $\Delta r = 0.5 \times 10^{-5}$ m are shown in Fig. 20a–g. The noise-like distributions of all decrypted results indicate that no valid information can be obtained. It is evident that the parameters of CSPM are remarkably sensitive decryption keys.

The CC values between the decrypted images and original images of R, G, and B channels of first image calculated with respect to the variation of x_0, p, k, m, λ, f, and r are plotted in Fig. 21a–g, respectively. It is evident that the CC values of R, G, and B channels reach one when all the correct parameters are applied for decryption. If the CC values deviate narrowly around their correct values, the CC curves decrease rapidly. From these plots, it is readily observed that a slight variation of parameters of CSPM will not allow the illegal deciphers to recover secret images.

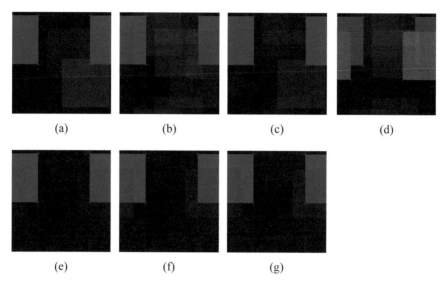

Fig. 20 Sensitivity of CSPM of the first image: **a** $\Delta x = 1 \times 10^{-16}$, **b** $\Delta k = 1$, **c** $\Delta p = 1 \times 10^{-15}$, **d** $\Delta m = 0.01$, **e** $\Delta \lambda = 0.1$ nm, **f** $\Delta f = 0.5 \times 10^{-4}$, **g** $\Delta r = 0.5 \times 10^{-5}$ m

5 General Conclusions

In this chapter, optical information cryptosystems based on structured phase encoding have been presented. The construction parameter(s) of structured phase mask offer(s) additional key(s) to the cryptosystem. The structured phase mask avoids problems arising from misalignment, which is an important issue in an optical system. Optical cryptosystems using radial Hilbert phase mask and Fresnel zone plate phase mask in the gyrator transform domain have been introduced in which construction parameter(s) is/are used as encryption key(s). Asymmetric information cryptosystems based on the spiral zone plate phase mask and optical interference principle, and chaotic spiral zone plate phase mask and random spectrum decomposition have been put forward in which construction parameters are used as decryption keys. Therefore, additional encryption/decryption keys guarantee the security of the designed optical single/multiple-color-image cryptosystems. The robustness and sensitiveness of parameters of structured phase mask as encryption/decryption keys remarkably enhance the security of the proposed techniques. Numerical simulation results have been presented to confirm the security, validity, and feasibility of the proposed cryptosystems.

Fig. 21 CC curve for deviation of **a** x_0, **b** p, **c** k, **d** m, **e** λ, **f** f, and **g** r

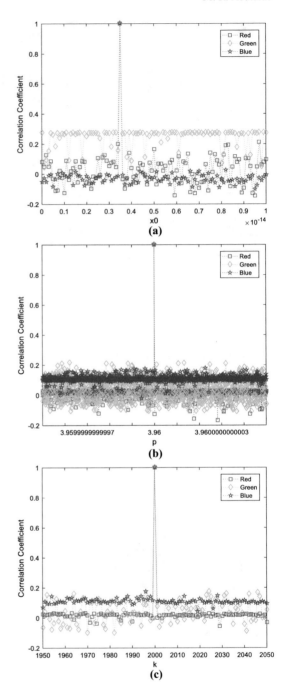

Fig. 21 (continued)

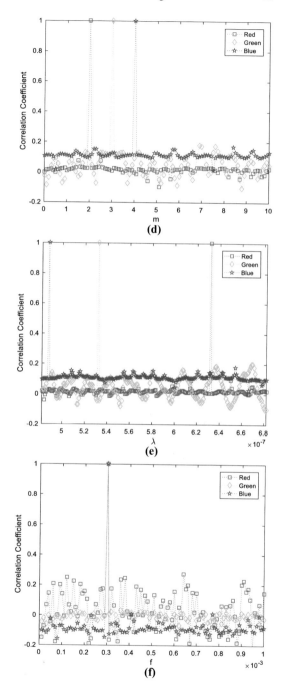

Fig. 21 (continued)

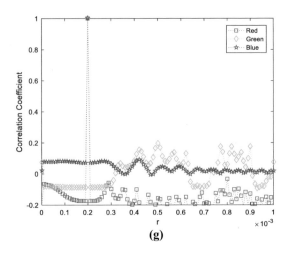

(g)

References

1. Alfalou A, Brosseau C (2009) Optical image compression and encryption methods. Adv. Opt. Photon. 1:589–636
2. Refregier P, Javidi B (1995) Optical image encryption based on input plane and Fourier plane random encoding. Opt Lett 20:767–769
3. Davis JA, McNamara DE, Cottrell DM, Campos J (2000) Image processing with the radial Hilbert transform: theory and experiments. Opt Lett 25:99–101
4. Aburabab MR (2012) Color information security system using discrete cosine transform in gyrator transform domain radial-Hilbert phase encoding. Opt Lasers Eng 50:1209–1216
5. Barrera JF, Henao R, Torroba R (2005) Optical encryption method using toroidal zone plates. Opt Commun 248:35–40
6. Aburabab MR (2012) Color image security system using double random-structured phase encoding in gyrator transform domain. Appl Opt 51:3006–3016
7. Aburabab MR (2012) Securing color image using discrete cosine transform in gyrator transform domain structured-phase encoding. Opt Lasers Eng 50:1383–1390
8. Sakdinawat A, Liu Y (2007) Soft-x-ray microscopy using spiral zone plates. Opt Lett 32:2635–2637
9. Aburabab MR (2013) Security enhancement of color image cryptosystem by optical interference principle and spiral phase encoding. Appl Opt 52:1555–1563
10. Sui L, Zhou B, Ning X, Tian A (2016) Optical multiple-image encryption based on the chaotic structured phase masks under the illumination of a vortex beam in the gyrator domain. Opt Express 24:499–515
11. Aburabab MR (2018) Asymmetric multiple information cryptosystem based on chaotic spiral phase mask and random spectrum decomposition". Opt Laser Technol 98:298–308
12. Rodrigo JA, Alieva T, Calvo ML (2007) Application of gyrator transform for image processing. Opt Commun 278:279–284
13. Rodrigo JA, Alieva T, Calvo ML (2007) Experimental implementation of the gyrator transform. J Opt Soc Am A 24:3135–3139
14. Zhang Y, Wang B (2008) Optical image encryption based on interference. Opt Lett 33:2443–2445
15. Wang Q (2012) Optical image encryption with silhouette removal based on interference and phase blend processing. Opt Commun 285:4294–4301

16. Qin W, Peng X (2010) Asymmetric cryptosystem based on phase-truncated Fourier transforms. Opt Lett 35:118–120
17. Cai J, Shen X, Lei M, Lin C, Dou S (2015) Asymmetric optical cryptosystem based on coherent superposition and equal modulus decomposition. Opt Lett 40:475–478
18. Wang Y, Quan C, Tay CJ (2016) New method of attack and security enhancement on an asymmetric cryptosystem based on equal modulus decomposition. Appl Opt 55:679–686
19. Li W-N, Phan A-H, Piao M-L, Kim N (2015) Multiple-image encryption based on triple interferences for flexibly decrypting high-quality images. Appl Opt 54:3273–3279
20. Li W-N, Shi C-X, Piao M-L, Kim N (2016) Multiple-3D-object secure information system based on phase shifting method and single interference. Appl Opt 55:4052–4059

Encryption/Decryption with Optical Transform

Zhengjun Liu, Yu Ji, Shurui Yang, Shutian Liu, Bin Gao, and Hang Chen

Abstract In this chapter, the methods of image encryption and decryption based on optical transformation are introduced. The extended fractional Fourier transform is constructed by designing an eccentric lens group, and is used in the image encryption system. By using the Fresnel diffraction model, an iterative phase recovery algorithm is constructed, and a color image hiding method and optical system are introduced. A parallel phase retreival technique in gyrator transform domain is constructed, and is applied to construct an image decryption system.

1 Introduction

Optical information security methods [1–8] have been innovated extensively, since the excellent advantages of optical information techniques, such as data processing with ultra-fast speed and multi-dimensions. Here, some optical transforms (fractional Fourier transform, gyrator transform, Fresnel transform) have a very important role for the strucuture of encryption system. In the visible wavelength range, the optical encryption scheme can be realized by the light field modulation of the microsystem. In this optical system, diffraction calculation with phase retrieval should be considered for secret image conversion for encryption. For recording data in optical systems, single-shot imaging can save time of information hiding, and color illumination and multiplexing technique were applied. Other optical information systems, such as diffraction imaging, interferometer, and joint transform correlator (JTC), were employed for encrypting secret picture. Multiple-image encryption has

Z. Liu (✉) · Y. Ji · S. Yang · S. Liu
School of Physics, Harbin Institute of Technology, Harbin 150001, China
e-mail: zjliu@hit.edu.cn

B. Gao
School of Data Science and Technology, Heilongjiang University, Harbin 150080, China

H. Chen
School of Space Information, Space Engineering University, Beijing 101416, China

been researched by ptychographic imaging, interference system, and digital holography. According to these encryption schemes, new optical process and transform system are beneficial to the research of optical information security field.

In this Chapter, a combination of two lenses is considered for controlling beam propagation carrying secret information in an information protection structure [9]. The two lenses are close together, but their center is not on the same optical axis. The lenses are embedded into the implementation system of an extended fractional Fourier transform (eFrFT). The mathematical model of the optical process is from the phase modulation and Fresnel diffractions. The physical size of the lenses can be regarded as several extra keys to enhance security. As the main body of keys, the random phase data is utilized for making random output images.

In the second part, a color image encryption method with the cascaded random phase modulation in the domains of Fresnel transform [10] is introduced. Its motivation is to seperate an original color image into the data of several phase masks. Generally, it should select two phase masks to encoding information in each channel for a simple system. Thereby, only four phase masks can implement for the task hiding a color image. Here, three phase images are regarded as the encryption output and combined into a color image, and the another phase pattern is selected as the key data of this encryption method. The three phase patterns are arranged in RGB chanels for Fresnel transforms. The phase image being the key is in the common channel of the encryption system. The iterative phase retrieval is applied for calculating the three phase pictures at monochrome channels. The real physical parameters of the image encryption system will serve as the additional cipher to hiding secret images.

In the third part, an iterative phase recovery algorithm [11] is introduced to solve the ill-conditioned problem of phase recovery. In this method, several images are obtained by changing the parameters of fractional transformation, which can be recorded as intensity patterns in the optical system. This algorithm is named as amplitude-phase retrieval (APR). In this phase recovery model, the amplitude and phase of the target to be measured are unknown. Compared with other phase recovery algorithms, the error of this algorithm is smaller. At the same time, with the increase of the number of captured images, the convergence speed of the algorithm is accelerated.

In the final part, a simple and effective double random phase coding (DRPE) in Fourier domain attack method is recalled [12]. The optical system includes point light source, pupil and CCD. Here CCD is used to record several diffracted light intensity patterns placed on the exit plane and plane outside the DRPE system along the optical axis, when the point light source illumination system on the input plane of the DRPE structure is adopted. A parallel iterative phase recovery algorithm based on Fresnel domain, is used to restore the phase distribution of the output plane of this optical encryption system. In DRPE system, the second phase mask data with little error can be obtained by inverse Fourier transform of the reconstructed complex amplitude function. This scheme is called light point attack. The phase distribution of the first random mask can also be measured and calculated, when the DRPE system is illuminated by a uniform beam.

2   Optical Image Encryption by Using eFrFT

2.1   Optical Encryption Scheme

In Fig. 1, the original image, $q_0(x_0, y_0)$ display on a spatial light modulator (SLM) is lighted by uniform beam from a beam splitter (BS), and is transmitted along the optical axis with the distance l_1 entering the random phase mask (PM). We can express the patern in the input plane of PM as follows

$$q_1(x_1, y_1) = \exp\left[i \cdot 2\pi \cdot r(x_1, y_1)\right]\mathcal{D}_{l_1, \lambda}\left[q_0(x_0, y_0)\right], \tag{1}$$

where $\mathcal{D}$ is an operator for Fresnel diffraction and the variable λ is wavelength of illumination beam. The function $r(x_1, y_1)$ denotes the phase information of PM, and is a uniform random data in the range [0, 1]. In the system in Fig. 1, eFrFT is implemented by a special structure of lenses. The random pattern modulated by the mask and the lenses is then finally received by CCD at the output end. In the system, the field function $q_2(x_2, y_2)$ is written as

$$q_2(x_2, y_2) = \exp\left[i \cdot t(x_2, y_2)\right]\mathcal{D}_{l_2, \lambda}\left[q_1(x_1, y_1)\right], \tag{2}$$

in which $t(x_2, y_2)$ is the phase function for the composite structure with two lenses. The phase function $t(x_2, y_2)$ is given as

$$t(x_2, y_2) = -\frac{\pi}{\lambda}\left[\frac{x_2^2 + y_2^2}{f_1} + \frac{(x_2 - x_s)^2 + (y_2 - y_s)^2}{f_2}\right], \tag{3}$$

where f_1 and f_2 are the focal length of two lenses. The variables (x_l, y_l) are used to express the centrifugal distance of two lenses. The phase modulation function of two lenses can also be realized by using a SLM to adjust the phase form digitally.

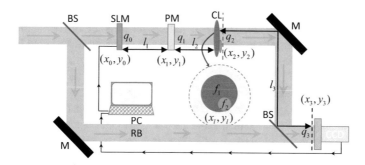

Fig. 1 The optical encryption system with the speial eFrFT by using a pair of convex lenses. BS: beam spliter; SLM: spatial light modulator; PM: phase mask; CL: combined lenses; M: mirror; RB: rererence beam; PC: computer

The pattern function $q_3(x_3, y_3)$ at the output plane can be received by CCD, and is computed by

$$q_3(x_3, y_3) = c_0 \left| D_{l_3, \lambda}[q_2(x_2, y_2)] \right|^2, \tag{4}$$

where c_0 is related to the response coefficient of CCD. The phase data matched with $q_3(x_3, y_3)$ can be measured by off-line holography. To enhance security, the data of q_3 can be converted by certain chaotic mapping before entering again into SLM via PC, for the operation of random phase encoding in the next loop of this encryption method. Therefore, these operations can be performed iteratively, to increase the degree of random of the final output pattern. For the simplicity of expression, the structure parameters (x_l, y_l) of lenses are taken at the unchanged value in the operations from all loops.

In the repeated optical information processing with this system, the distance values l_1, l_2 and l_3, are also fixed for simplifying experimental operations. These five values matched with the position of optical elements, and two values of focal distance (f_1 and f_2) can be regarded as some additional passwords to improve the security of this encryption algorithm. The reverse optical structure of the system shown in Fig. 1 can be applied for image decryption, where conjugate element and inverse operation will be adopted, such as the conjugate phase mask and two concave lenses (with the focal lengths, $-f_1$ and $-f_2$).

2.2 Numerical Simulation

For validating the performance of this optical encryption method, some numerical simulation results are displayed, here. The simulation parameters are selected as: (a) $\lambda = 632.8$ nm, $f_1 = 10$ cm, $f_2 = 3$ cm, $x_l = 2.3$ mm, $y_l = 2$ mm, $l_1 = 6$ cm, $l_2 = 5$ cm, $l_3 = 11$ cm, (b) the radius of small lens is set at 1.6 mm, (c) the physical size of original input pattern is 1 cm × 1 cm. The encryption operation is performed 6 times iteratively. A gray-level image having 300×300 pixels serves as original secret image, and is drawn in Fig. 2a. The phase pattern information of PM is illustrated in Fig. 2b. The amplitude and phase of final encryption results are plotted in Fig. 2c, d. These output pictures in Fig. 2c, d have demonstrated that this image encryption method is effective to protect the secret visual information. The secret patterns can be transformed completely into orignal status, when all correct values of keys are used in the decryption computation. The corresponding pattern is given in Fig. 3. These calculated results have shown that this encryption scheme is run successfully. Here Arnold mapping is used for pixel scrambling, and its period is 600 for the pixel size of input image.

The security from all keys will be tested and discussed. First, the main key $p(x_1, y_1)$ of this encryption method is checked. A wrong phase $p'(x_1, y_1)$ is obtained by introducing a normally distributed noisy data with mean value of 0 and standard

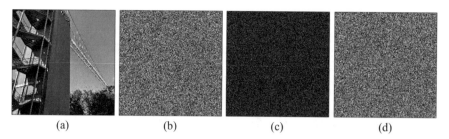

(a)   (b)   (c)   (d)

Fig. 2 The simulation result of encryption: **a** original pattern, **b** random phase for PM, **c** amplitude pattern of output field, **d** phase pattern of output field

Fig. 3 The results of image decryption

deviation of 0.08, to be added into the true data of the phase function $p(x_1, y_1)$. In DRPE, the phase key is used to make a comparison with the system in Fig. 1. The phase $p(x_1, y_1)$ is adopted in the proposed method and the key data of DRPE. Two decrypted images are calculated by the changed phase $p'(x_1, y_1)$, and are shown in Fig. 4. Here other keys of new method are fixed as right values in this test. The peak signal to noise ratio (PSNR) function is selected for weighting the errors of retrieved images. The results implied that this method (PSNR = 10.16 dB) is safer than DRPE (PSNR = 15.57 dB) under the protection of the phase key $p(x_1, y_1)$.

The decryption test is calculated by use of some values of two focal distances (f_1 and f_2) of two lenses. Here, one focal length value uses the correct value, while the other focal length value is incorrect. In this test, values of other keys are correct. The range of scanning for focal length, is achieved from 2 to 12 cm, which has the right values of two parameters, f_1 and f_2. The PSNR of all decrypted images are curved in Fig. 5. From the curves, the parameters from focal length are very sensitive to blind evaluating a key for any illegal user. To increase the number of keys can linearly improve the security of the encryption algorithm. Moreover, these keys cooperate with each other, to ensure the security of secret images.

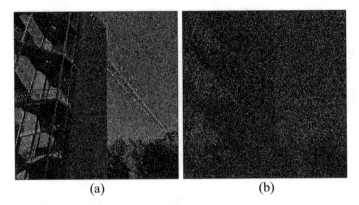

Fig. 4 The results of decryption with incorrect phase $p'(x_1, y_1)$: **a** DRPE with PSNR = 15.57 dB, **b** the proposed method with PSNR = 10.16 dB

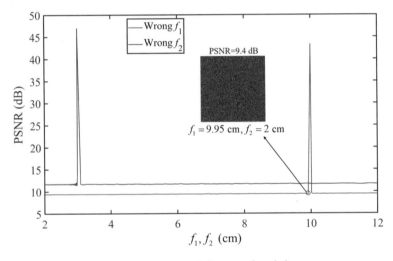

Fig. 5 The decryption errors with incorrect focal distances f_1 and f_2

The central offset (x_l, y_l) of the lenses in eFrFT structure are also treated as a pair of additional keys. The PSNR distribution is calculated in the range [1 mm, 3 mm], and is given in Fig. 6. Here exists a very vimineous peak being at the position defined by true values. The results in Fig. 6 make clear that the right values of variables x_s and y_s ensures the quality of decrypted images and provides good security. The restored image in Fig. 6 is extremely blurred visually, and it contains only the faint outline information of the original image. Here x_s is 2.2 mm. The variable y_s and other keys are taken at right values.

The sensitive situation of distances (d_1, d_2, d_3), and wavelength λ, is checked individually by using blind scanning calculation. Four PSNR curves are computed and shown in Fig. 7. The retrieved images have high precision according to the

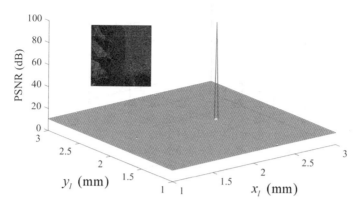

Fig. 6 The PSNR distribution from wrong values of (x'_s, y'_s). The decrypted image is generated with $x'_s = 2.2$ mm and $y'_s = 2$ mm, and its PSNR equals 12.2 dB

PSNR values when the key parameters are right. These results demonstrate that the wavelength λ is subtler than three distance parameters when its value is traversed. Moreover the sensitive case of distance parameters can be increased, if the variation is added simultaneously into the variables, d_2 and d_3, which are the length of two arms of optical system implementing eFrFT.

Similarly, the wavelength of the illuminating beam, λ, and the distance between optical elements, (l_1, l_2, l_3), are also the single-value keys of this encryption method.

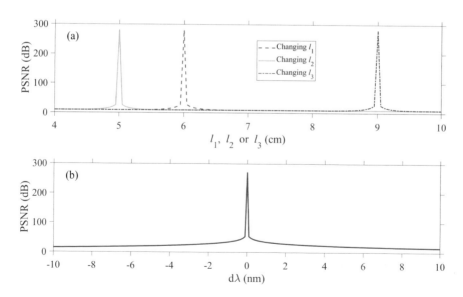

Fig. 7 The error curves of changed distances and wavelength: **a** distance, **b** wavelength

The PSNR curves of the response, when their values are changed is shown in Fig. 7. These curves show that the spacing and wavelength are very sensitive to the response, when deviating from the exact values. They are feasible as passwords in the application. These parameters are from the main parameters of extended fractional Fourier transform, to define the optical strycture of transform.

Although PSNR function is very effective to evaluate the error of two images. However, for the image, with the change of the total energy of the optical system, the pattern meaning can still be accurately identified for computer vision. In this case, the PSNR function excitation will not be prepared, and inappropriate conclusions will be given when evaluating the quality of the decrypted image by using PSNR.

Here, we consider using the correlation coefficient to evaluate the similarity between the two images (original image I_0 and retrieved image I_r). Its mathematical definition is as follows

$$E[I_0] = \frac{1}{MN} \sum_{\forall m,n} I_0(m,n),$$

$$D[I_0] = \frac{1}{MN} \sum_{\forall m,n} \{I_0(m,n) - E[I_0(m,n)]\}^2,$$

$$\text{cov}[I_0, I_r] = \frac{1}{MN} \sum_{\forall m,n} \{I_0(m,n) - E[I_0]\} \times \{I_r(m,n) - E[I_r]\},$$

$$\text{cc}[I_0, I_r] = \frac{|\text{cov}[I_0, I_r]|}{\sqrt{D[I_0(m,n)] \cdot D[I_r(m,n)]}},$$

(5)

where M and N are the pixel sizes. The variables E and D are mean value and variance, for the matrix of input images. CC can express the similarity between two checked images having a relation $I_r = cI_0$, which c is single-value parameter and is from total intensity change of illumination beam. The evaluated image I_r has a better quality, if CC value is equal to 1.

Occlusion attacks are considered and used to simulate part of the data loss of an encrypted image. Here it is considered that the data in the upper-left corner of the encrypted image is missing. The missing data is replaced by a value of 0 in the calculation. The corresponding test results are shown in Fig. 8. The CC value of the decrypted image here is equal to 0.85. From Fig. 8b, the main information of the original image is recognizable in this case.

The multiplicative noise model is used to simulate noise attack experiments. It is defined as

$$q_3'(x, y) = q_3(x, y)[1 + t \cdot G(x, y)],$$

(6)

where q_3' and q_3 are the polluted and the encrypted patterns. The parameter t is to determine the intensity of the noise function G. The matrix G has standard deviation 1 and mean value 0. In the calculation, in order to give more statistical results, several different values of matrix G are calculated here. At the same time, the corresponding

(a)   (b)

Fig. 8 The test result of image occlusion: **a** occluded pattern, **b** decoded secret image

CC value is calculated and drawn into a curve. Here the maximum value of the parameter t is set to 1.5. The corresponding curve is shown in Fig. 9. It can be seen from the figure that as the value of t increases, the CC values decrease accordingly, that is, the quality of the decrypted image becomes worse. In order to visually monitor the image quality, a decrypted image is calculated and given, where the parameter t is equal to 0.6.

Fig. 9 The CC curve under noise test. Here CC value of decrypted image is 0.66

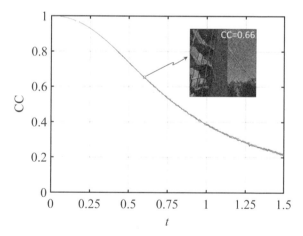

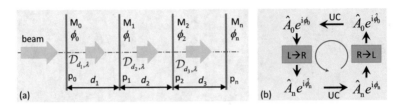

Fig. 10 The model of cascaded phase-only encoding: **a** optical structure, **b** phase retrieval. UC: updating constraint

3 Image Encryption by Using Phase-Only Encoding

3.1 Phase Encoding

In view of the visual invisibility of optical phase coding, it is considered that the secret image is hidden in the phase information. For this reason, a cascaded phase coding structure is considered, as shown in Fig. 10. Several phase plates are arranged on the optical axis in turn. The conversion relationship between them can be expressed by Fresnel diffraction. The phase of the input and output planes in the system can be preset to secret images and encrypted images. The middle phase plates can compensate and balance the optical system through them. According to the analysis, at least one phase board is needed to balance the phase conversion relationship in the system.

The iterative algorithm structure can be used for the phase recovery problem within the structure in Fig. 10a, and the calculation process is shown in Fig. 10b. The numerical calculation of the beam propagation process can be completed by Fresnel diffraction. Here $R \to L$ and $L \to R$ represent the transmission calculation of the beams on the left and right sides. The phase distribution functions on the input and output planes can be used as supporting constraints in the phase retrieval algorithm to update the calculated phase distribution. The amplitude distribution here is free, which can be used to balance the phase distribution of each plane. The phase plates on the planes of P_1 and P_2 can be used for phase compensation on both sides.

3.2 Proposed Encryption System

According to the previous analysis, the optical encryption system shown in Fig. 11, can be designed. The RGB color lighting optical system is used here. Fresnel diffraction is applied to calculate the light field propagation between phase plates. The beams of the three monochromatic channels are summarized by the beam splitter and diffracted through the phase plate M_c for a certain distance to form a secret image. Here the secret image is presented in the form of intensity. Its corresponding complex-valued light field can be expressed as

$$C_{0,n}(x, y) = \sqrt{I_{0,n}(x, y)} \exp[i \cdot 2\pi\varphi_n(x, y)], \ (n = r, g, b), \tag{7}$$

where $I_{0,n}$ denotes intensity functions. The index n is the color of light. According to the calculation method given in Fig. 10, the phase distribution function φ_n of the plane, where the phase plate M_c is located, can be obtained.

In order to increase the modulation randomness of the phase plate, the phase plate M_c in this optical system can be rotated and translated. The light field on the exit plane of the phase plate M_c can be expressed as

$$U_n(x_1, y_1) = \mathcal{D}_{-d, \lambda_n}\big[C_{0,n}(x, y)\big], \tag{8}$$

where the value of negative distance, $-d$, can obtain the result of inverse diffraction in numerical calculation. λ_n is wavelength of the corresponding beams. The phase function ϕ_n can be generated by

$$S_{0,n} \exp[i \cdot \phi_n(x_0, y_0)] = \mathcal{D}_{-l_n, \lambda_n}\big\{U_n(x_1, y_1) \exp[i \cdot \phi_c(x_1, y_1)]\big\}. \tag{9}$$

Here the phase ϕ_n is used to adjust the beam phase distribution of the three channels in the system. The distance l_c gives the corresponding value of optical path. The function $S_{0,n}$ can be used to balance the energy conservation of the input and output planes. The value of amplitude function $S_{0,n}$ can be uniformly distributed or a function with intensity fluctuation. For the beams of uniform intensity, the value $S_{0,n}$, can be calculated according to the following formula

$$S_{0,n} = \sqrt{\frac{1}{\sigma_v} \iint_\sigma I_{0,n}(x, y)\mathrm{d}x\mathrm{d}y}, \tag{10}$$

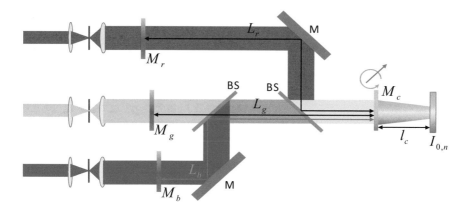

Fig. 11 The information security system to encrypt color images. BS: beam splitter; M_r, M_g, M_b, M_c: designed phase masks

where σ_v and σ represent the area size and location of the region in which the image I_0 is located. In Eq. 9, the four phase functions are calculated, and the secret image information is embedded in these phases. The phase ϕ_n can be calculated according to the iterative phase recovery algorithm. The pattern of phase ϕ_n is random, which is suitable for applications of image encryption. The main structural parameters of the encryption system will be determined, when the four phase functions are calculated. In addition, wavelength λ_n, optical path l_c and the parameters of geometric operations (rotation and shifting) of phase mask M_c can be used as additional ciphers, to improve the security of secret images.

In this encryption algorithm, the secret image I_0 is converted into four phase functions by the phase recovery algorithm. In view of the fact that the phase plate M_c is placed in the main optical path, three branch optical paths can be controlled by ϕ_c simultaneously. This phase ϕ_c, is more suitable as a key to improve security.

In this encryption structure, the limitation of amplitude distribution is not taken into account. If the amplitude is encoded into other secret images, more intermediate phases need to be taken into account to balance the difference between the amplitude and phase of the transition plane. The corresponding advantage is that more secret images can be hidden, while the corresponding inverse problem calculation needs to be considered more, and the amount of computation is increased. At the same time, the hardware implementation of the optical system is much more difficult.

The security of the corresponding algorithm can be further improved, if more phase plates are added to modulate the beam in Fig. 11. As an application form of deformation, this algorithm can be used as a form of multi-image encryption. The effect of crosstalk noise in multi-image encryption can be avoided by encoding images with 3 channels.

3.3 Numerical Simulation

Here, some numerical simulation results are given to verify the effectiveness of the color image encryption algorithm and the related performance tests. The wavelength and optical path parameters used in the calculation are listed in Table 1, and the control parameters of the geometric operation of the phase plate in the main optical path are also given. The number of calculations for iterative phase retrieval is set to 2000.

Table 1 The physical parameters in the encryption system

Quantity	Value (nm)	Quantity	Value (m)	Quantity	Value
λ_b	633	L_r	1.5	d	0.5 m
λ_g	532	L_g	0.9	θ	$\pi/6$
λ_b	470	L_b	1.3	$(\Delta x, \Delta y)$	(3, 2)px

A secret color image having 256×256 pixels (px) is selected and given in Fig. 12a. Physically, the image has a size of 2 cm $\times$ 2 cm. A noise-like pattern with 512×512 px displayed in Fig. 12b is used for mask M_c. In order to select a range of pixels from the phase mask M_c to match the secret image, bilinear interpolation is used to generate a small matrix of data from a large matrix. The red box in Fig. 12 gives an example of a selection. The position of the red box is obtained by rotation and translation. The random phase distribution patterns of the corresponding colors are given in Fig. 12. Their graphic contents are random and can be used to encrypt the results, and to store and transmit them in the application. The three random and composite color random patterns are shown in Fig. 12.

To test the security provided by the keys, this consideration changes the keys and uses it to decrypt the image. Firstly, the noise pollution of the phase ϕ_c in the main optical path is considered. Here, a two-dimensional matrix data is added to the phase data, the mean value is 0 and the variance is σ_0. The corresponding decryption result is shown in Fig. 13. The correlation coefficient of the corresponding image is listed in Table 2. The image in Fig. 13a is restored using a completely correct password, although the correlation value is not strictly equal to 1 (mainly because the iterative phase recovery algorithm is difficult to absolutely converge to the true value), but its

| (a) | (b) | (c) |
| (d) | (e) | (f) |

Fig. 12 The encryption results: **a** color secret image, **b** phase image ϕ_c, **c** phase ϕ_r, **d** phase ϕ_g, **e** phase ϕ_b, and **f** color synthetic image

Table 2 Correlation coefficient values in Fig. 13

Images	CC_r	CC_g	CC_b
Figure 13a	0.981	0.990	0.993
Figure 13b	0.700	0.736	0.757
Figure 13c	0.285	0.304	0.333

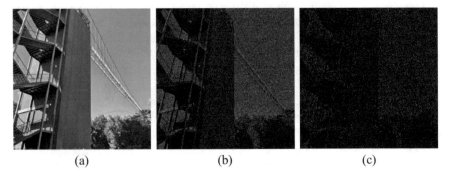

(a)   (b)   (c)

Fig. 13 The images obtained with the wrong values of phase key ϕ_c: **a** correct key, **b** the noise σ = 0.1, **c** the noise σ = 0.25

visual effect is very close to the original image. Moreover, $\sigma_0 = 0.25$ is regarded as an evaluated threshold for checking quality of decoded images.

When the distance between the wavelength and the phase plate changes, the corresponding decryption process is also calculated by ergodicity. The corresponding error curve is shown in Fig. 14. The error variables are added to the true values of the corresponding physical quantities respectively. The decryption pattern of 0.002 nm and 0.002 mm when the wavelength and spacing occur respectively is also given in the figure. It is also difficult to see the outline information of the original image visually. It shows that they are very sensitive to very small changes. They can bring good security to encryption algorithms.

The monochromatic components of the decryption pattern are decrypted when there is a small error between a set of wavelengths and phase plate spacing given in Fig. 15. As can be seen from the results, the decrypted image is very sensitive to the changes of these two physical quantities, which will be very difficult for illegal users to decrypt blindly.

In Fig. 16, the geometric operation of the phase plate M_c is analyzed, including rotation and translation. Here the standard deviation (STD) case is used to describe the phase transformation of phase ϕ_c because it is related to the quality of the decrypted image. Therefore, the standard deviation transformation corresponding to translation and rotation is drawn in Fig. 16. The vertical axis of a curve or surface represents the standard deviation. The models of rotation and translation are described by the following formula

$$\theta' = \theta + \delta\theta, \ \Delta x' = \Delta x + \delta x, \ \Delta y' = \Delta y + \delta y \tag{11}$$

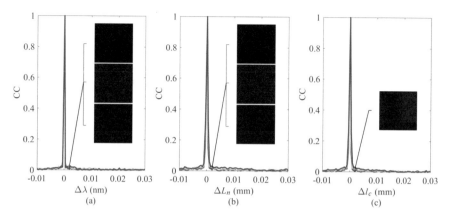

Fig. 14 The results by the changed parameters $\Delta\lambda_n$, ΔL_n and Δl_c: **a** wavelength λ_n, **b** lengthes L_n, and **c** distance l_c. The curve color is coincident with monochrome images

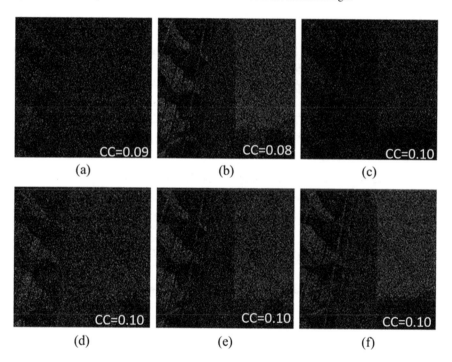

Fig. 15 The decrypted images generated by the values incorrect wavelength and optical path: **a** $\Delta\lambda_r = 2.0 \times 10^{-4}$ nm, **b** $\Delta\lambda_g = 4.5 \times 10^{-4}$ nm, **c** $\Delta\lambda_b = 2.5 \times 10^{-4}$ nm, **d** $\Delta L_r = 4.6 \times 10^{-4}$ mm, **e** $\Delta L_g = 7.2 \times 10^{-4}$ mm, **f** $\Delta L_b = 6.9 \times 10^{-4}$ mm

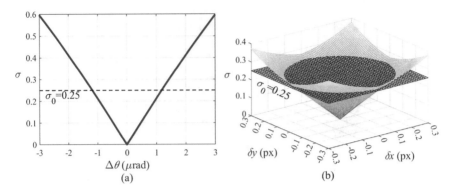

Fig. 16 The test results of geometric operation of the phase mask M_c: **a** $\Delta\theta$, **b** δx and δy. The dash line and mesh maked with the threshold $\sigma_0 = 0.25$ are the boundaries of image quality

where θ' and $(\Delta x', \Delta y')$ are wrong values and are applied in decryption for θ and $(\Delta x, \Delta y)$.

Here the STD of phase 0.25 is used as the threshold to monitor the image quality, and the corresponding image quality can be determined according to Fig. 13c. A range below the threshold will be considered a valid decryption range.

4   Phase-Amplitude Retrieval in Gyrator Domains

Here, a phase retrieval algorithm based on fractional transforms is introduced. The transformation used is selected as the gyrator transform.

4.1   Optical System and Phase Retrieval Scheme

The mathematical formula of gyrator transform can be expressed as

$$Q(u, v) = G^\alpha[q(x, y)]$$
$$= \frac{1}{|\sin\alpha|} \iint q(x, y) \times \exp\left[i2\pi\frac{(xy + uv)\cos\alpha - (xv + yu)}{\sin\alpha}\right]dxdy \tag{12}$$

where Q and q represent the output and input functions of the transform. The parameter α of the transformation is the order of the fractional transform. This transform can be achieved using six cylindrical lenses, which are divided into three groups, each with two cylindrical lenses. The parameter α is related to the angle between two cylindrical lenses. Three groups of cylindrical lenses are arranged along the optical axis in turn. By changing the value of the order α, some different transformed patterns can be obtained for imaging and measurement.

The calculation process of this phase recovery algorithm is shown in Fig. 17. Some transformation patterns I_n, can be recorded by different fractional orders. Mathematically, there are the following relation

$$I_n(u, v) = [A_n(u, v)]^2 = |G^{\alpha_n}\{A_0(x, y)\exp[i\phi_0(x, y)]\}|^2, \qquad (13)$$

where A_n is an amplitude function from fractional α_n. These amplitude functions will be used as supporting constraints in the phase retrieval algorithm. Here, the amplitude and phase functions of the target are unknown. They need to be obtained by the phase retrieval algorithm. Therefore, this algorithm is called as the amplitude and phase retrieval (APR). The main idea of this method is to operate the mean on the target plane to be tested.

In the iterative calculation as Fig. 17, the initial values of the amplitude and phase of the target to be measured are set to random values. Then through the calculation of different fractional order iterations in the spatial domain and frequency domain. As the iteration goes on, the estimated values of amplitude and phase will quickly approach the true value. When the number of measured images increases, the convergence of the corresponding algorithm will become better.

A set of comparative results of phase recovery, including convergence curve and error distribution, are given in Fig. 18. The fractional order used in the calculation is listed in Table 3.

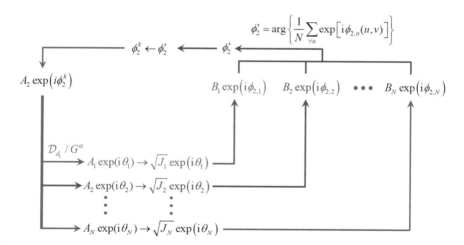

Fig. 17 The retrieval scheme of phase and amplitude

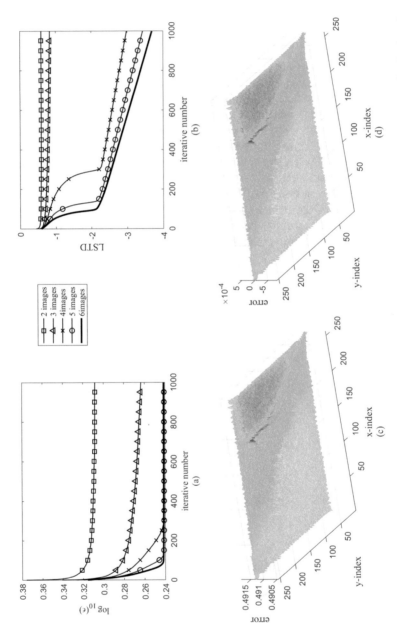

Fig. 18 The comparison of convergence for two kinds of retrieval method. **a** Error mean value, **b** logarithm of STD (LSTD), **c** error distribution of mean value, **d** without dc component

Table 3 The values of transform orders in APR

Test	α_1	α_2	α_3	α_4	α_5	α_6
2 images	0.7	1.0	/	/	/	/
3 images	0.7	1.0	1.2	/	/	/
4 images	0.7	1.0	1.2	1.35	/	/
5 images	0.7	1.0	1.2	1.35	1.4	/
6 images	0.7	1.0	1.2	1.35	1.4	1.42

4.2 Analysis and Discussion

In the APR method, all strengths on the output plane (the information in the input is not required) are used as supporting constraints. The intensity image is calculated in parallel by fractional transform and its inverse transform. Updating the light field of the incident plane by using the arithmetic mean of the complex amplitude is helpful to suppress the noise in the experimental image. This method can be used for intensity and phase measurement in some cases, such as computational imaging and quantitative phase imaging.

It needs to be added here, that when analyzing the phase error, the coherent optical system is not sensitive to the constant phase difference. Therefore, it is more practical to consider removing the DC component or using the standard deviation to evaluate the phase error.

5 Phase-Only Encoding in Fresnel Domains

With the help of the previous APR method, this section considers attack testing for double random phase coding (DRPE). The encryption system consists of a 4f optical system. We consider to record multiple diffraction spots at the output and use them for phase recovery [12].

5.1 Optical System and Scheme of Attack on DRPE

DRPE implemented by the 4f system is shown in Fig. 19. Its mathematical description can be expressed as

$$
\begin{aligned}
E_1(u, v) &= \mathcal{F}\{s(x, y) \exp[i\varphi_1(x, y)]\}, \\
E_2(x', y') &= \mathcal{F}\{E_1(u, v) \exp[i\varphi_2(u, v)]\},
\end{aligned}
\tag{14}
$$

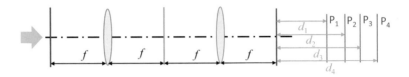

Fig. 19 An optical system to attack DRPE. A pupil implements a support constraint for the light field in the output plane of DRPE

where E_1 and E_2 are the output of the first and second lenses. $s(x, y)$ represents the function corresponding to the secret image. φ_n is random phase functions. Here $\mathcal{F}$ is Fourier operator. This random phase coding process is realized by two-step Fourier transform. For image cracking, if the attacker steals the distribution data of the second phase board, then the encryption system will become very dangerous. The secret images are transparent to attackers. Therefore, the key to image cracking is to calculate the phase distribution information of the second phase plate.

A point light source lighting system is used in Fig. 19. It can be prepared by pupil filtering to filter the focused spot on the front focal plane of the first lens. Mathematically, the pulse function is used to describe the point light source. In this case, the function E_2 is

$$\begin{aligned} E_2(x', y') &= \exp[i\varphi_1(0, 0)]\mathcal{F}\{\exp[i\varphi_2(u, v)]\} \\ &= A_2(x, y)\exp[i\phi_2(x', y')], \end{aligned} \tag{15}$$

where the amplitude image A_2 can be obtained by calculated from the measured intensity pattern at the output plane of DRPE in Fig. 19. Specially, The phase distribution information of the first phase plate is not related to decryption. This conclusion is also consistent with the practical application of DRPE. Other methods need to be considered in order to calculate the phase distribution information of the second phase. No disassembly measurement of the DRPE system is considered here. It is only measured from outside the system. In order to measure the light field on the output plane of DRPE system, off-axis holography can be used. But this technology requires a more complex system. Here, the coherent diffraction imaging scheme is considered to measure the light field information of the output plane.

The APR method of Fresnel transform domain is considered here. As shown in Fig. 19, we can use CCD to move along the optical axis and record the Fresnel diffraction spot at different positions, for marking it as $I_{2,n}$. These spots can be modeled using Fresnel diffraction, that is,

$$I_{2,n}(x_n, y_n) = c_0\left|\mathcal{D}_{d_n}\left[E_2(x', y')\right]\right|^2, \quad n = 1, 2, ..., N, \tag{16}$$

where N is the total number of measured images. c_0 is a constant, that balances the total energy of the light field between the planes. The distance from dn is the distance from the corresponding recording plane to the DRPE output plane. In addition, the light field intensity of the output surface can be directly recorded by CCD, and the amplitude distribution function of the output surface can be obtained by squaring it. In order to accelerate the convergence of the phase recovery algorithm, a pupil can be set on the output surface to limit the measurement range. At the same time, measuring the amplitude distribution of the output surface is also helpful to accelerate the convergence accuracy of the phase recovery algorithm. Compared with the previously introduced APR algorithm in the gyrator transform domain, the calculation target here is only the phase information of the light field E_2, and its amplitude information does not need to be calculated.

In order to solve the phase of E_2, the calculation process of the corresponding phase recovery algorithm can be referred to Fig. 17. The negative distance can be used to calculate inverse Fresnel diffraction. The phase can be updated by calculating the mean of the complex value, that is,

$$\phi_2' = \arg\left\{\frac{1}{N}\sum_{\forall n}\exp\left[i\phi_{2,n}(u, v)\right]\right\}, \tag{17}$$

where the average phase ϕ_2' is used as the updated values of phase ϕ_2^k in the calculation of next loop. The function "arg" is to compute the angle of input complex values. After running the APR method, the phase distribution information of E_2 can be obtained, and the corresponding light field can be written as $\hat{E}_2(x', y')$. The phase information of the second phase plate φ_2, in DRPE method can be calculated according to the following formula

$$\hat{\varphi}_2(u, v) = \arg\left\{\mathcal{F}^{-1}\left[\hat{E}_2(x', y')\right]\right\}, \tag{18}$$

where the function $\hat{\varphi}_2$ represents a relative phase distribution. The absolute phase distribution cannot be obtained here because the origin phase of the first phase plate in Eq. 15 cannot be determined by measurement and calculation. At the same time, there is also a DC phase in the phase retrieval algorithm. In view of the fact that this constant error does not affect image decryption, its effect is ignored here.

This attack method is named as PLSA because it uses a point light source to complete the attack measurement. This method requires the use of a point light source, CCD and pupil. Few components are used and the structure of the system is simple. The DRPE system is placed in the inner space of this attack method. The beam emitted by the point light source passes through the DRPE system and a plurality of diffraction spots are recorded by the CCD. The iterative phase recovery algorithm can calculate the password of DRPE, that is, the phase data of the second phase plate. For the information of the first phase plate, the DRPE system can be illuminated by a uniform intensity laser beam, and the data of the phase φ_1 can be obtained by using the previously measured phase $\hat{\varphi}_2$ via inverse Fourier transform.

5.2 *Numerical Result and Discussion*

In the calculation, the data scale of secret image and random phase is 256×256 px. For optical structure, the width and height of the used image are taken at 1 cm. The wavelength of light beam is 632.8 nm. The pattern obtained by the DRPE method is shown in Fig. 20a. In order to accelerate the convergence of phase retrieval, it can consider adding a circle of zero elements around the random pattern. Its pixel width is w as Fig. 20b. The values of distance in calculation is listed in Table 4.

The standard deviation of error (STDE) between the true phase ϕ_2 and recovered phase ϕ_2^k are calculated as

$$\text{STDE} = \text{std}(\Delta\phi) = \text{std}(\phi_2 - \phi_2^k), \qquad (24)$$

where the function std is to compute standard deviation for input matrix. STDE can remove the interference of DC component to phase error evaluation. Several corresponding STDE curves are given in Fig. 21. Here the operation of common logarithm (base 10) is applied to check the convergence status of iterations for small values of error. The curve consists of three stages: slow decline, rapid decline, and

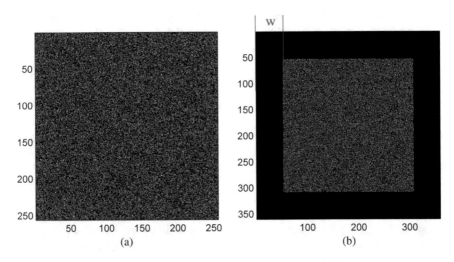

Fig. 20 An encryption result with point light source regarding as a secret pattern. **a** amplitude, **b** support constraint. Here, w = 52 px

Table 4 Values of distance d_n in diffraction (Unit: m)

Distance	d_1	d_2	d_3	d_4	d_5	d_6
$N = 4$	0.1	0.15	0.2	0.25	/	/
$N = 5$	0.1	0.15	0.2	0.25	0.3	/
$N = 6$	0.1	0.15	0.2	0.25	0.3	0.35

steady state. The fast convergence phase will be obtained easier, if more measurement patterns are used. As can be seen from Fig. 21, STDE reaches a very small value, which means that the error of phase retrieval is very small.

The error distribution of the phase distribution calculated after 1000 iterations is shown in Fig. 22. From the error distribution results, it can see that the phase accuracy of the phase retrieval algorithm result is very high.

In order to test the effect of recovery phase in image decryption, an 8-bit grayscale image, Baboon, containing 256×256 px, is used. The corresponding results are shown in Fig. 23.

Here, the intensity images generated by six point light sources are used to calculate the phase retrieval. The number of iterations is 180, 185 and 190. Three decrypted images are shown in Fig. 23b–d. The image in Fig. 23d has almost no visual noise, which is very close to the original image. 200 iterations are sufficient for restoring secret images, when six measurement images are used. Figure 23e shows the mean square error (MSE) curve between the restored image and the original image. The

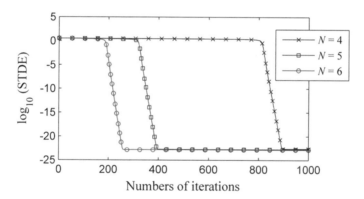

Fig. 21 Convergence curves of the phase retrieval algorithm with different intensity patterns. Here the error is generated with the recovered phase and its true value for the second phase mask

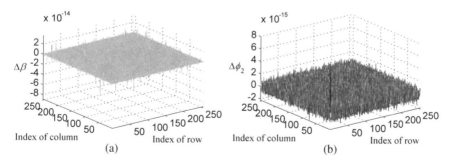

Fig. 22 Error distribution of recovered phases. **a** β_2^k and **b** $\hat{\phi}_2$. The maximum values of relative error of $\Delta\beta$ and $\Delta\phi_2$ are 8.86×10^{-14} and 8.02×10^{-15}, respectively

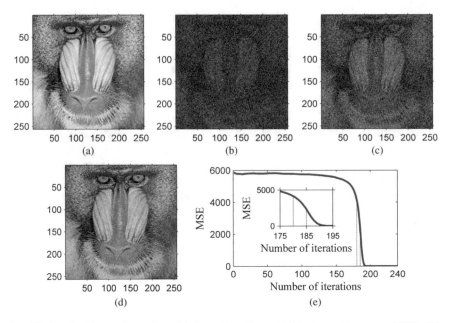

Fig. 23 Result of image decryption with the retrieved key. **a** Original secret image. **b–d** With 180, 185, and 190 iterations. **e** MSE curve

MSE value of the decrypted image is less than 0.0259. Figure 23 proves that the PLSA scheme can successfully decrypt the DRPE system.

6   General Conclusions

In this chapter, the image encryption scheme of iterative random phase coding in eFrFT domains is introduced. Optical transform is realized by a structure composed of two eccentric lenses. The structural parameters of the optical system, such as distance, wavelength and mismatch distance at the center of the lenses, can be used as additional keys. The use of these specially designed keys improves the security of the encryption algorithm.

Secondly, a color image hiding scheme using pure phase coding in Fresnel transform domain is introduced. The color secret image is distributed as the output amplitude of the encryption system. Independent phase keys are placed on the common channel to modulate red, green and blue beams. Three pure phase masks are arranged at the input of the designed optical system. The phase distributions of three pure phase masks are obtained by using the phase recovery algorithm. The three phase-only masks contain all the information of the original image. The secret image will be optically reconstructed, when the tricolor uniform beam illuminates the system.

Thirdly, APR method in gyrator transform domain is introduced to reconstruct the intensity and phase of the target. Several measured intensity images are obtained from the output plane by using different order gyrator transform, and they are used as supporting constraints for data reconstruction. The information of the target light field is unknown. The algorithm has advantages in suppressing noise.

Finally, a simple and effective attack scheme of DRPE system is introduced. The optical elements used include point light source, pupil and CCD. The parallel iterative phase recovery algorithm is used to reconstruct the phase distribution of the output plane of the encryption system. After the estimated phase diagram is obtained, the relative phase of the second mask of DRPE is calculated by inverse Fourier transform. This scheme can restore the secret image with a small error.

Acknowledgements This work was supported by the National Natural Science Foundation of China (Nos. 11874132, 61975044, 12074094), Interdisciplinary Research Foundation of HIT (No. IR2021237).

References

1. Refregier P, Javidi B (1995) Optical image encryption based on input plane and Fourier plane random encoding. Opt Lett 20:767–769
2. Peng X, Zhang P, Wei H, Yu B (2006) Known-plaintext attack on optical encryption based on double random phase keys. Opt Lett 31:1044–1046
3. Peng X, Wei H, Zhang P (2006) Chosen-plaintext attack on lensless double-random phase encoding in the Fresnel domain. Opt Lett 31:3261–3263
4. Yamaguchi I, Zhang T (1997) Phase-shifting digital holography. Opt Lett 22:1268–1270
5. Gerchberg RW, Saxton WO (1972) A practical algorithm for the determination of phase from image and diffraction plane pictures. Optik 35:237–246
6. Liu Z, Xu L, Lin C, Liu S (2010) Image encryption by encoding with a nonuniform optical beam in gyrator transform domains. Appl Opt 49:5632–5637
7. Yang B, Liu Z, Wang B, Zhang Y, Liu S (2011) Optical stream-cipher-like system for image encryption based on Michelson interferometer. Opt Express 19:2634–2642
8. Rodrigo JA, Duadi H, Alieva T, Zalevsky Z (2010) 2010 Multi-stage phase retrieval algorithm based upon the gyrator transform. Opt Express 18:1510–1520
9. Liu Z, Chen H, Blondel W, Shen Z, Liu S (2018) Image security based on iterative random phase encoding in expanded fractional Fourier transform domains. Opt Lasers Eng 105(1):1–5
10. Liu Z, Guo C, Tan J, Liu W, Wu J, Wu Q, Pan L, Liu S (2015) Securing color image by using phase-only encoding in fresnel domains. Opt Lasers Eng 68:87–92
11. Liu Z, Guo C, Tan J, Wu Q, Pan L, Liu S (2015) Iterative phase-amplitude retrieval from multiple images in gyrator domains. J Opt 17:025701
12. Liu Z, Shen C, Tan J, Liu S (2016) A recovery method of double random phase encoding system with a parallel phase retrieval. IEEE Photon J 8(1):7801807

Optical Cryptosystems Based on Spiral Phase Modulation

Ravi Kumar, Yi Xiong, and Sakshi

Abstract Optical information security techniques have several advantages over digital counterparts such as ability to process information parallelly, use of physical parameters as security keys, efficient storage capability etc. In last few years, several optical cryptosystems have been designed based on different optical aspects. In this chapter, we discuss optical cryptosystems based on spiral/vortex phase modulation in details. The orbital angular momentum (OAM) associated with a spatially helical phase or vortex beam can be utilized to design enhanced security protocols. Moreover, since the OAM has theoretically unlimited values of topological charges (TCs) and have the orthogonality of OAM modes with different integer TCs, it is an excellent candidate for designing high-capacity secure optical cryptosystems. Here, the spiral phase functions have been first introduced with different TCs and then the 2D spiral phase transform (SPT) and several optical cryptosystems based on it are discussed in detail with possible optical configurations for practice applications. Numerical simulation results for three cryptosystems are discussed showing their feasibility. The security analysis in terms of keys sensitivity and robustness against existing attacks is also performed and discussed for these cryptosystems.

R. Kumar (✉)
School of Electrical and Computer Engineering, Ben-Gurion University of the Negev, P.O. Box 653, 8410501 Beer-Sheva, Israel
e-mail: ry20724@gmail.com

Y. Xiong
School of Science, Jiangsu Provincial Research Center of Light Industrial Optoelectronic Engineering and Technology, Jiangnan University, Wuxi 214122, China

Sakshi
Department of Chemical Engineering, Ben Gurion University of the Negev, Beer-Sheva, Israel

R. Kumar
Department of Physics, SRM University - AP, Andhra Pradesh, Amaravati 522540, India

1 Introduction

Nowadays, due to vast use of technology in the world has prompted several challenges for safety of data and its secured processing. There are several instances where the data transmitted through public networks has been breached and misused for illegal activities. Thus, in current times the development of enhanced cryptosystems is very critical and inevitable. Recently, optical encryption methods have gained a lot of popularity due to their inherent advantages over digital counterparts such as, parallel processing, numerous degrees of freedoms (i.e. amplitude, phase, polarization, wavelength etc.), fast computing speed, multidimensionality [1, 2]. The first optical cryptosystem, double random phase encoding (DRPE) was reported back in 1995 by Réfrégier and Javidi [3]. In DRPE, a two-dimensional (2D) images is encrypted into a white noise like image by modulating the original information using random phase masks in spatial and frequency domain. A 4-f optical setup was employed to achieve the optical encryption in DRPE [3]. With time, several modifications were made in the original DRPE technique to improve the security by using different linear transforms, such as fractional Fourier [4], Fresnel transform [5], Gyrator transform [6], and Mellin transforms [7, 8]. However, it was found that the DRPE-based techniques are inherently linear and vulnerable to various attacks such as chosen-plaintext attack (CPA) [9], chosen-cyphertext attack (CCA) [10], and known-plaintext attack (KPA) [11]. A new technique based on DRPE architecture is also reported using hybrid phase masks which is robust against KPA and additional security with large key space [12]. In order to improve the security, some other aspects of optics have also been explored [13–18]. Optical cryptosystems based on joint transform correlator [13], Shearlet transform [14], wavelength and position multiplexing [15, 16], non-separable linear canonical transforms [17], and diffractive imaging [18] have been introduced for security enhancement. Further, various mathematical decomposition such as equal/unequal modulus decomposition [19–21], polar decomposition [22], and singular value decomposition [23] has also been integrated with optical cryptosystems which significantly improved the security. Some latest encryption cryptosystems using the spiral phase modulation have been reported [24–26]. In these techniques, instead of modulating the wavefront with random phase masks, the spiral phase modulation is utilized. The robustness of these cryptosystems is demonstrated against several types of attacks.

In this chapter, optical cryptosystems based on spiral/vortex phase modulation are discussed in detail. The orbital angular momentum (OAM) associated with a spatially helical phase or vortex beam brings a new degree of freedom which can be utilized for carrying information with enhanced security [27–30]. Basically, the spiral phase in the vortex beam has various locations where the phase is uncertain, it may have a value of 0 or 1. These points where the phase has undefined values are called the singular points given by the topological charges (TCs) (q) [24]. Moreover, since the OAM has theoretically unlimited values of TCs and the orthogonality of OAM modes with different integer TCs [31], it is an excellent candidate for designing high-capacity secure optical cryptosystems. This chapter starts with the brief introduction of the spiral phase functions, then the 2D spiral phase transform (SPT) and several optical cryptosystems based on it are discussed in details.

The rest of the chapter is organized as following. In Sect. 1, the general introduction of spiral phase function is given. Then three different optical cryptosystems based on spiral phase transform are summarized in Sect. 2. In Sect. 3, the concluding remarks are given.

2 Spiral Phase Transform (SPT)

The Spiral phase transform (SPT) was introduced for demodulation of 2D fringe patterns [32]. It is obtained by using a 2D signum function, sgn (u, v), also called spiral phase function (SPF). The plot of half-plane signum function as a function of (u, v) is shown in Fig. 1 and it is used for 2-D Hilbert transform [32, 33] in order to calculate SPT mathematically.

The 2-D signum function can be defined in the spatial frequency space (u, v) as [32]:

$$SPF = \text{sgn}(u, v) = \frac{u + iv}{\sqrt{u^2 + v^2}} = \exp\{i\phi(u, v)\} \tag{1}$$

where the phase, $\phi(u, v)$ is the polar angle in frequency space which gives information about the coordinate rotation in the polar coordinate system. It is to be noted that the SPF function is undefined at the origin, at which it may have values either zero or one, which is called the singularity point in the phase function [33].

For optical cryptosystems, the number of singularity or undefined points in the phase function are utilized as the security key. The number of singular points can be varied in SPF by modifying it by defining a parameter, q called the topological charge of the SPF. The modified SPF can be given as [24]:

$$MSPF = \exp\{iq\phi(u, v)\} \tag{2}$$

Fig. 1 Schematic illustration of half-plane signum function as a function of frequency coordinates

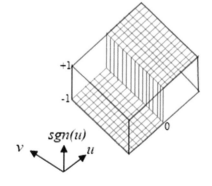

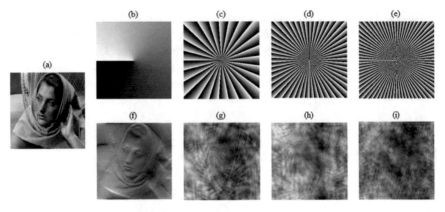

Fig. 2 (**a**) Original input image, (**b–e**) SPFs with $q = 1, 25, 50$, and 75, respectively, (**f–i**) the corresponding SPTs of (**a**) with $q = 1, 25, 50$, and 75, respectively

Now using this, the SPT can be defined by modulating the input function in the Fourier domain. For a 2D function $f(x, y)$, the SPT can be given as [24]:

$$SPT\{f(x, y)\} = IFT\{MSPF.FT\{f(x, y)\}\} \tag{3}$$

where the operators FT and IFT are the forward and inverse 2-D Fourier transforms, respectively. The inverse SPT can be calculated by using the complex conjugate of MSPF as:

$$ISPT\{f(x, y)\} = IFT\{conj(MSPF).FT\{f(x, y)\}\} \tag{4}$$

where $conj(.)$ is a function which gives the complex conjugate of input. Figure 2 depicts the MSPF with various values of topological charge (q) and the SPTs of an input image with changing q.

3 Optical Cryptosystems Based Spiral Phase Modulation

The spiral phase function modulates the incoming wavefront by introducing the singular phase points at multiple location. The number of singular points is given by the TC. For practical applications, a spatial light modulator (SLM) can be used to modulate the light in an optical cryptosystem. The value of TC plays a significant role while designing the optical cryptosystem based on spiral phase modulation. Furthermore, complex structured phase masks having very high randomness can also be generated using MSPF by introducing lateral, quadrature and other shifts. Now, in the upcoming subsections, three optical cryptosystems [24–26] utilizing MSPF are discussed with simulation results and their possible optical configuration.

3.1 Nonlinear Optical Cryptosystem Based on SPT in Fresnel Domain

In this section, a nonlinear optical image encryption technique using SPT in Fresnel domain is discussed. To introduce the nonlinearity to the cryptosystem, a power function is used [24]. The original 2D image is first transformed to a phase image and then modulated with a random amplitude mask (RAM) to get the complex image. RAM has random values between 0 and 1. Then the power function with nonlinearity parameter, m, is used on the complex image which returns an image with powered values of each pixel. This powered image is then optical propagated with Fresnel diffraction to a distance z_1. The resultant wavefront after at a distance z_1 is then modulated with a random phase mask (RPM) using SLM and Fresnel propagated with distance z_2 is recorded. Similarly, in a separated channel a security image is Fresnel propagated with distance z_3 after modulating with another RAM. Then, the two complex images obtained using image and secondary image after above mentioned process, are combined to get an intermediate complex image. This complex image is further Fresnel propagated to a distance z_4 and finally, the SPT with predefined MSPF is performed to get the final encrypted image. The schematic flowchart of the encryption process is shown in Fig. 3.

For optical implementation of this encryption technique an optoelectronic setup illustrated in Fig. 4 can be used. SLMs are employed to display the RPM and RAMs which modulate the incoming wavefront. The MSPFs shown in Fig. 1b–d are displayed on SLM_3 to perform the SPT with predefined TC. SLM_3 is placed at the focal plane of two Fourier lenses, L_1 and L_2, here we have used the two lenses having same focal lengths, for example, 200 mm. The final encrypted image is recorded by a charge coupled device (CCD) controlled using the personal computer (PC).

For validation, a series of numerical simulations were carried out using MATLAB™. The results are presented in Fig. 5. The original input image and security image are shown in Fig. 5a and b, respectively. Figure 5c and d show the complex

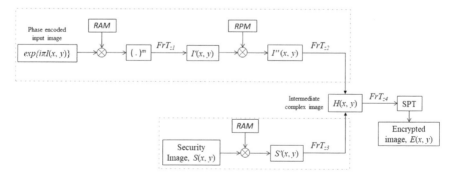

Fig. 3 Schematic of encryption process. RAM, random amplitude masks; m, nonlinearity parameter; RPM, random phase mask; and FrT_z, Fresnel propagation with distance z. Figure adapted from [24]

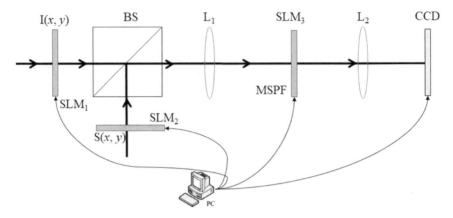

Fig. 4 Proposed optoelectronic setup for practical implementation. Figure adapted from [24]

images after modulation with RAM and Fresnel propagation of input and security image, respectively. The intermediated complex image is shown in Fig. 5e, whereas the final encrypted image is shown in Fig. 5f.

The presented encryption method has a large key space including, nonlinearity parameter, MSPF order, RPM, Fresnel propagation distances etc. The sensitivity and applicability of all the security keys is checked by performing the decryption process with correct and incorrect keys. Figure 6 shows the results for decryption.

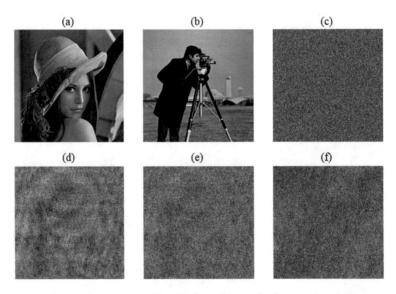

Fig. 5 Encryption results: (**a**) original input image, (**b**) security image, (**c**) and (**d**) the complex images after modulation with RAM and Fresnel propagation of (**a**) and (**b**), respectively, (**e**) intermediated complex image, and (**f**) final encrypted image. Figure adapted from [24]

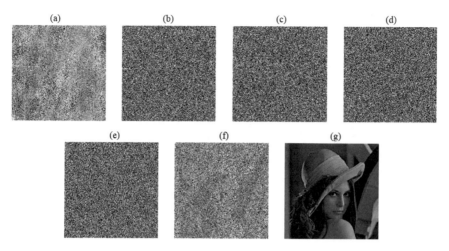

Fig. 6 Decrypted results. The decoded image with: (**a**) wrong z_1, (**b**) wrong z_2, (**c**) wrong z_4, (**d**) wrong MSPF order, (**e**) wrong RPM, (**f**) without using the security image steps, and (**g**) all the correct keys. Figure adapted from [24]

The decrypted images with wrong z_1, z_2 and z_4, are shown in Fig. 6a–c, respectively, whereas, Fig. 4d and Fig. 6e are the decrypted images with incorrect MSPF order, and different RPM, M_1, respectively. When the steps involving security image, $S(x, y)$ are not used the decrypted image is shown in Fig. 6f. The final decrypted image with all the correct keys is shown in Fig. 4g. To further check the effectiveness normalized mean square error (NMSE) is also calculated between the original and decoded image in all the cases. When the decryption is performed with even one wrong key, the NMSE values become very large. For the decrypted images shown in Fig. 6a–f the NMSE values were 0.1983, 0.4241, 0.4207, 0.4198, 0.4212 and 0.2735, respectively. These large NMSE values confirms the effectiveness of each key and its significance in terms of enhanced security. It is to be noted that when all the correct keys are used for decryption, the NMSE value was very small (NMSE = 0.0099).

Furthermore, the security of the discussed technique is also tested against noise contamination and brute force attack. The corresponding results are depicted in Fig. 7. The encrypted image is contaminated with Gaussian noise of two different strengths as shown in Fig. 7a and b. Then the original image is decoded from these contaminated ciphertexts, i.e. Fig. 7c and d. From the results, it can be seen that the decoded images from the contaminated ciphertexts also reveals most of the information and the content of the original image can be easily recognized visually. Figure 7e shows the NMSE values for 1000 different RPMs used in place of the original RPM in the proposed scheme. The value of NMSE was always more than 0.4, which confirms that no valuable information can be retrieved with wrong RPM and the technique is robust against such brute force attack.

We have further analyzed the effect of nonlinearity parameter, m and noticed that for all values of m between $[-1, -3]$ and $[1, 4.5]$, the presented technique works

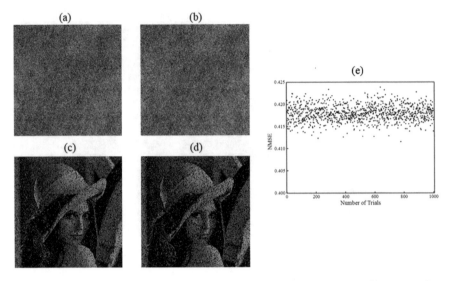

Fig. 7 Attack analysis: (**a**) and (**b**) noise contaminated encrypted images, (**c**) and (**d**) corresponding decrypted images using (**a**) and (**b**), respectively, (**e**) brute force attack results. Figure adapted from [24]

well, but for values of m outside these ranges, the recovered image has some saturated pixels. This can be attributed to the singularities present in the MSPF.

3.2 Multiuser Optical Authentication Using Photon Counting in SPT Domain

In this section, an SPT based optical information authentication technique is discussed. To design the system, photon counting imaging (PCI) approach which is also utilized along with polar decomposition which makes the scheme asymmetric [25]. In polar decomposition (PD), two private keys are generated which enhances the security of the system and enables the multiuser platform [22]. Only, one of these two private keys are needed at a given time. Thus, these keys can be sent to two different users for simultaneous access to the information. For authentication, a nonlinear correlation is utilized which has better correlation peaks even with less sparse samples. In PCI, the number of incident photons, N_p can be controlled in the image recording process [34–36]. This results into a photon-limited image having less information than the original image. The photon counting probability, lj at a given pixel x_j can be calculated by Poisson-distribution. In the present scheme, PCI is used on the complex image obtained from SPT. The flowchart of the encryption and authentication processes is given in Fig. 8.

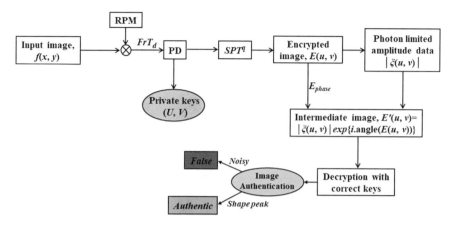

Fig. 8 Flowchart of optical information encryption and authentication technique. Figure adapted from [25]

During encryption, an input image, $f(x, y)$ as input image is first modulated with a RPM and then Fresnel propagated with distance d, to get the complex image. Afterwards, the PD operation is applied on the complex image, which decompose it into three matrices, i.e. two symmetric matrices and a rotational matrix. The symmetric matrices will be used as the private keys for authentication and the SPT with MSPF is performed on the rotational matrix, obtain a complex image or primary encrypted image, $E(u, v)$. Now using the PCI technique, the photon-limited amplitude, $|\xi(u, v)|$ of the encrypted image is generated. It is noteworthy to mention that; the phase of the encrypted image is reserved for decryption or authentication and only the amplitude is used in the PCI. Then the photon limited amplitude obtained after PCI and the phase of encrypted image are combined to synthesize an intermediate image, $E'(u, v)$ which is then used for authentication. Usually, the decrypted images from the PCI based techniques are not visually recognizable as these techniques are mostly mostly used for verification or authentication [25]. In authentication, the decrypted image, $d(x, y)$ is compared with the original image $f(x, y)$ using a nonlinear correlation (NLC) [35]. The inverse Fourier-transform of their product in frequency domain will give the nonlinear correlation between both the images.

To validate and check the feasibility of the method, numerical simulations are performed using MATLAB™ on a laptop with Intel core i5 processer and 8 GB RAM. The simulation results are shown in Fig. 9. An input image, shown in Fig. 9a is used for the validation. Figure 9b and c shows two private keys obtained from the PD operation. The amplitude and phase part of the complex image after SPT are shown in Fig. 9d and e, respectively. The photon-limited amplitude from PCI, $|\xi(u, v)|$ with $N_p = 2 \times 10^4$ is shown in Fig. 9f.

Fig. 9 Simulation results: (**a**) input image, (**b**) and (**c**) private keys after PD, respectively, (**d**) amplitude of encrypted image $E(u, v)$, (**e**) phase of encrypted image $E(u, v)$, (**f**) photon-limited amplitude, $|\xi(u, v)|$ with $N_p = 2 \times 10^4$. Figure adapted from [25]

The decrypted image with same number of photons and all the other correct keys is shown in Fig. 10a and the corresponding NLC plot is given in Fig. 10b. The Fresnel propagation distance of $d = 50$ mm and MSPF with TC, $q = 20$ was used for the simulations. The NLC operation is performed between the decrypted image and the input image (taken as reference). The sharp peak in the NLC plot (Fig. 10b) confirms the high degree of correlation and validates the applicability of the presented method. We have also analyzed the sensitivity of different security keys in the cryptosystem. Figure 10c, e and g show the recovered images when wrong Fresnel distance, d, (changed by 10 mm), wrong SPT order, q (changed by 4), and wrong private key, respectively, whereas, the corresponding NLC plots are given in Fig. 3d, f and h. When any one of the security key parameters is incorrect or slightly changed there is no sharp peak in the NLC plot which indicates the week correlation and confirms that no information about the original image is present [34]. Thus, for true authentication all the correct keys must be used. Also, it is to be noted that anyone of the two private keys from PD may be used for true authentication enabling the multiuser platform which is an additional advantage.

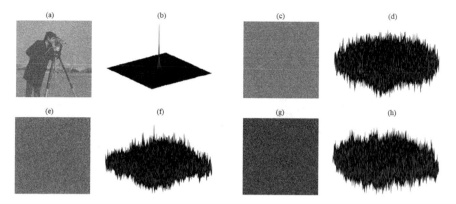

Fig. 10 Decrypted image with photon-limited amplitude: (**a**) all correct keys, (**c**) wrong Fresnel distance (*d*), (**e**) wrong SPT order, (**g**) wrong private key, (**b**), (**d**), (**f**) and (**h**) are the corresponding NLC plots for (**a**), (**c**), (**e**), and (**g**), respectively, with reference image shown in Fig. 9a. Figure adapted from [25]

3.3   Optical Cryptosystem for Color Image Encryption Based on SPT and Chaotic Pixel Scrambling

Here, a SPT-based optical cryptosystem for color image encryption is discussed [26]. To enhance the security chaotic pixel scrambling based on Tinkerbell map is used along with the structured phase masks (SPM) [37]. Tinkerbell map is a chaotic 2D discrete-time dynamical system [38]. It used five initial parameters to generate two chaotic sequences which are used for pixel scrambling. For shuffling of image pixels, first, the chaotic sequences must have elements equal to the dimensions of the input image, i.e. $M \times N \times 3$ for the color image. Two examples of scrambling a gray scale and binary image using Tinkerbell map along with its attractor are shown in Fig. 11. The initial values used for the chaotic map are -0.72, 0.64, 0.9, -0.6, 2, and 0.5.

For encryption, SPMs with different parameters are used. First, the color image is separated into red, green, and blue components and each of these are modulated individually with the SPMs. The modulated RGB components are then used as inputs to the SPT with a specific order of MSPF. The complex outputs after SPT are then decomposed into two parts through the random modulus decomposition (RMD) technique. The one part after RMD is stored as the private key and the other part is scrambled using Tinkerbell map to get the encrypted image. All the three encrypted images corresponding to RGB components are combined to get the final color encrypted image. The block diagram of the color cryptosystem is shown in Fig. 12.

The encryption/decryption processes of the technique can be performed with both digital and optical processing. For optical realization, a combination of two SLMs, Fourier lenses (L), Laser, and CCD is used. The schematic of the optoelectronic systems is shown in Fig. 13. The three RGB components are processed individually. SPMs are displayed on the SLMs to modulate the coming wavefront. A 4-f

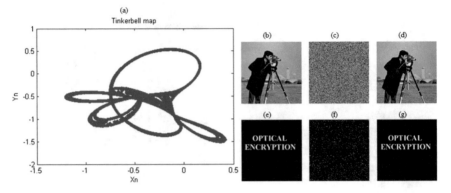

Fig. 11 (**a**) Attractor of Tinkerbell map, (**b**) and (**e**) grayscale and binary input image, respectively, (**c**) and (**f**) the scrambled images using Tinkerbell map, (**d**) and (**g**) the recovered images, respectively. Figure adapted from [26]

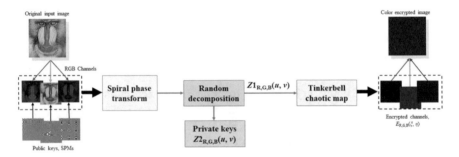

Fig. 12 Block diagram of the color optical cryptosystem. Figure adapted from [26]

optical setup with two Fourier lenses is utilized to perform the SPT by using a MSPF displayed on the SLM in the Fourier plane.

Numerical simulation experiments were performed to validate the technique using MATLAB™ (version R2016b) on a PC having an Intel(R) Xeon(R) CPU E5-2630 v4 processor and 32 GB RAM. A color image of 'Flowers', shown in Fig. 14a is used as the original input image. Figure 14b–d, show the three RGB components of the input color image, whereas, the SPMs used for modulation are shown in Fig. 14e–g, respectively. The individual RGB encrypted images are shown in Fig. 14h–j, whereas, Fig. 14k shows the final encrypted color image.

A detailed analysis is also performed to check the effectiveness of security keys. Figure 15a–c show the private keys generated during encryption process of each R, G, and B components, respectively. When all the correct security keys, i.e. three SPMs, SPT order, private keys and correct de-scrambling with Tinkerbell map, are used for decryption the recovered image is shown in Fig. 15d. Further, when the one of the security keys is either wrong or not used while decoding process no information is revealed, as shown in Fig. 15e–h. Moreover, the robustness of the technique is also

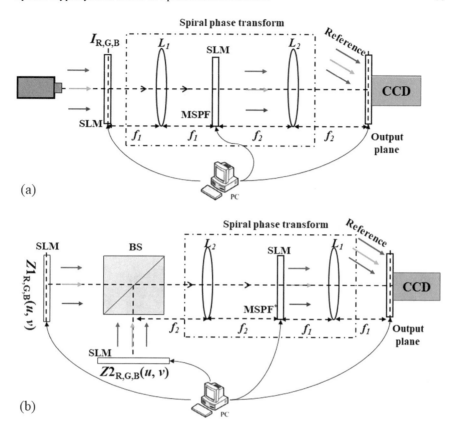

Fig. 13 Optoelectronic setup for experiments: (**a**) encryption process, and (**b**) decryption process. Figure adapted from [26]

checked against various attacks, i.e. noise and occlusion attack, plaintext attacks, and special iterative attacks. The corresponding results are shown in Fig. 16. For noise attack, the ciphertext is contaminated with a Gaussian noise with strength 0.4. The contaminate ciphertext is shown in Fig. 16a and the corresponding decrypted image is shown in Fig. 16b. For occlusion test, some pixels (15%) of the encrypted image are replaced with value zero and then the decryption is performed with this occluded ciphertext. The corresponding results are shown in Fig. 16c and d. The various attacks, i.e. KPA and CPA are also performed on the presented technique and the results are shown in Fig. 16e and f, respectively. Lastly, a special iterative attack is also performed on the ciphertext and the plot of correlation coefficient (CC) values are calculated in each iteration. The plot of CC values with number of iterations is given in Fig. 16g along with a decrypted image after 500 iterations in the inset. The results confirm the validity, effectiveness, and robustness of the presented technique.

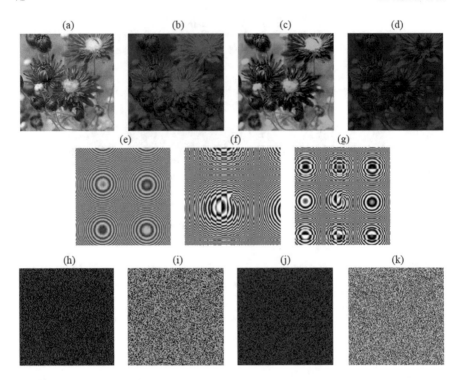

Fig. 14 Simulation results for encryption. (**a**) Original color image, (**b**)–(**d**) R, G, B components, respectively, (**e**)–(**g**) SPMs used for modulation, (**h**)–(**j**) R, G, B encrypted components, and (**k**) final color encrypted image. Figure adapted from [26]

4   Concluding Remarks

In this chapter, the spiral phase transform and its applications in the field of optical encryption have been discussed in details. Three different optical cryptosystems using combination of SPT with other optical aspects are elaborated along with their opto-electronic setups. The validity and capability of the spiral phase modulation and its advantages has been demonstrated by various kinds of numerical simulations results. According to the results, the security of the traditional cryptosystem is improved due to the presence of singular or vortex points in in the MSPF. The effectiveness of the MSPFs has been demonstrated for linear, nonlinear encryption along with the application to color image encryption. In future, the OAM properties with SPF can also be explored in order to design more sophisticated optical cryptosystems.

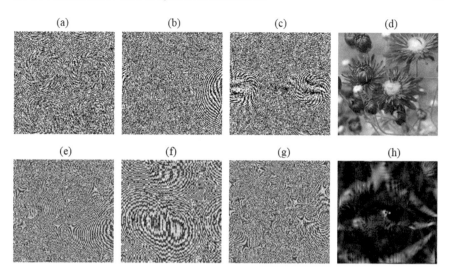

Fig. 15 Simulation results for decryption and key sensitivity analysis. (**a**)–(**c**) Private keys for R, G, and B components, respectively. Decrypted image: (**d**) with all correct keys, (**e**) when chaotic de-scrambling is not used, (**f**) without private keys, (**g**) wrong private keys, (**h**) wrong TC value in MSPF. Figure adapted from [26]

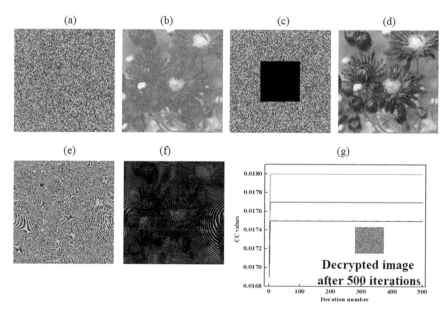

Fig. 16 Attack analysis results: (**a**) noise contaminated ciphertext, (**b**) decrypted image from (**a**), (**c**) ciphertext with occlusion, (**d**) decrypted image form (**c**), (**e**) and (**f**) the recovered image from KPA and CPA, respectively, (**g**) results of special iterative attacks. Figure adapted from [26]

References

1. Chen W, Javidi B, Chen X (2014) Advances in optical security systems. Adv Opt Photon 6:55–120
2. Nischal NK (2019) Optical cryptosystems. Institute of Physics (IOP) Publishing Ltd, 437 Bristol, UK. https://doi.org/10.1088/978-0-7503-2220-1
3. Refregier P, Javidi B (1995) Optical image encryption based on input plane and Fourier plane random encoding. Opt Lett 7:767–769
4. Unnikrishnan G, Joseph J, Singh K (2000) Optical encryption by double-random phase encoding in the fractional Fourier domain. Opt Lett 25:887–889
5. Matoba O, Javidi B (1999) Encrypted optical memory system using three-dimensional keys in the Fresnel domain. Opt Lett 24:762–764
6. Liu Z, Xu L, Lin C, Liu S (2010) Image encryption by encoding with a non-uniform optical beam in gyrator transforms domains. Appl Opt 49:5632–5637
7. Deng X, Zhao D (2011) Single-channel color image encryption using a modified 497 Gerchberg-Saxton algorithm and mutual encoding in the Fresnel domain. Appl Opt 50:6019–6025
8. Zhou N, Wang Y, Gong L (2011) Novel optical image encryption scheme based on fractional Mellin transform. Opt Commun 13:3234–3242
9. Peng X, Wei H, Zhang P (2006) Chosen-plaintext attack on lensless double-random phase encoding in the Fresnel domain. Opt Lett 31:3261–3263
10. Carnicer A et al (2005) Vulnerability to chosen-cyphertext attacks of optical encryption schemes based on double random phase keys. Opt Lett 30:1644–1646
11. Gopinathan U et al (2006) A known-plaintext heuristic attack on the Fourier plane encryption algorithm. Opt Express 14:3181–3186
12. Kumar R, Bhaduri B (2017) Optical image encryption using Kronecker product and hybrid phase masks. Opt Laser Technol 95:51–55
13. Barrera JF, Tebaldi M, Ríos C, Rueda E, Bolognini N, Torroba R (2012) Experimental multiplexing of encrypted movies using a JTC architecture. Opt Express 20:3388–3393
14. Kumar R, Bhaduri B, Hennelly B (2018) QR code based nonlinear image encryption using Shearlet transform and spiral phase transform. J Mod Opt 65:321–330
15. Amaya D, Tebaldi M, Torroba M, Bolognini N (2009) Wavelength multiplexing encryption using joint transform correlator architecture. Appl Opt 48:2099–2104
16. Qin Y, Gong Q (2013) Interference-based multiple-image encryption with silhouette removal by position multiplexing. Appl Opt 52:3987–3992
17. Kumar R, Bhaduri B, Sheridan JT (2018) Nonlinear double image encryption using 2D non-separable linear canonical transform and phase retrieval algorithm. Opt Laser Technol 107:353–360
18. Chen W, Chen XD, Sheppard C (2010) Optical image encryption based on diffractive imaging. Opt Lett 35:3817–4389
19. Kumar R, Bhaduri B, Quan C (2017) Asymmetric optical image encryption using Kolmogorov phase screens and equal modulus decomposition. Opt Eng 56:113109
20. Sachin KR, Singh P (2021) Unequal modulus decomposition and modified Gerchberg Saxton algorithm based asymmetric cryptosystem in Chirp-Z transform domain. Opt Quant Electron 53:254
21. Luan G, Wang D, Huang C (2020) Asymmetric image encryption and authentication in interference-based scheme using random modulus decomposition. J Mod Opt 67:1379–1387
22. Kumar R, Quan C (2019) Asymmetric multi-user optical cryptosystem based on polar decomposition and Shearlet transform. Opt Lasers Eng 120:118–126
23. Tao S, Tang C, Shen Y, Lei Z (2020) Optical image encryption based on biometric keys and singular value decomposition. Appl Opt 59:2422–2430
24. Kumar R, Bhaduri B (2017) Optical image encryption in Fresnel domain using spiral phase Transform. J Opt 19:095701

25. Kumar R, Quan C (2019) Multiuser optical information authentication using photon counting in spiral phase transform domain. Proc SPIE 11027:110270D. https://doi.org/10.1117/12.252 0844
26. Kumar R, Quan C (2019) Optical color image encryption using spiral phase transform and chaotic pixel scrambling. J Mod Opt 66:776–785
27. Luo X et al (2020) Integrated metasurfaces with microprints and helicity-multiplexed holograms for real-time optical encryption. Adv Opt Mater 8:1902020
28. Yao A, Padgett MJ (2011) Orbital angular momentum: origins, behavior and applications. Adv Opt Photon 3:161–204
29. Shen Y et al (2019) Optical vortices 30 years on: OAM manipulation from topological charge to multiple singularities. Light Sci Appli 8:90
30. Fang X, Ren H, Gu M (2020) Orbital angular momentum holography for high-security encryption. Nat Photonics 14:102–108
31. Allen L, Beijersbergen MW, Spreeuw RJC, Woerdman JP (1992) Orbital angular momentum of light and the transformation of Laguerre-Gaussian laser modes. Phys Rev A 45:8185–8189
32. Larkin KG, Bone DJ, Oldfield MA (2001) Natural demodulation of two-dimensional fringe patterns. I. General background of the spiral phase quadrature transform. J Opt Soc Am A 18:1862–1870
33. Lu Y, Li R, Lu R (2016) Fast demodulation of pattern images by spiral phase transform in structured-illumination reflectance imaging for detection of bruises in apples. Comp Electron Agric 127:652–658
34. Wang Y, Markman A, Quan C, Javidi B (2016) Double-random-phase encryption with photon counting for image authentication using only the amplitude of the encrypted image. J Opt Soc Am A 33:2158–2165
35. Perez-Cabre E, Abril HC, Mill MS, Javidi B (2012) Photon-counting double-random-phase encoding for secure image verification and retrieval. J Opt 14:094001
36. Rawat N, Hwang IC, Shi Y, Lee BG (2015) Optical image encryption via photon-counting imaging and compressive sensing based ptychography. J Opt 17:065704
37. Kumar R, Bhaduri B, Nischal NK (2018) Nonlinear QR code based optical image encryption using spiral phase transform, equal modulus decomposition and singular value decomposition. J Opt 20:015701
38. Goldsztejn A, Hayes W, Collins P (2011) Tinkerbell is Chaotic. SIAM J Appl Dyn Syst 10:1480–1501

Image Cryptosystem for Different Kinds of Image by Using Improved Arnold Map

Hang Chen, Yanhua Cao, Shutian Liu, Zhengjun Liu, and Zhonglin Yang

Abstract In this chapter, the methods of image encryption and decryption based on optical transformation are introduced. The extended fractional Fourier transform is constructed by designing an eccentric lens group, and is used in the image encryption system. By using the Fresnel diffraction model, an iterative phase recovery algorithm is constructed, and a color image hiding method and optical system are introduced. A parallel phase retrieval technique in gyrator transform domain is constructed, and is applied to construct an image decryption system.

1 Introduction

When it comes to image information security protection, digital image encryption is the most commonly used and the most important information processing technology [1–3]. It has important academic significance and application value for the protection of personal privacy, trade secrets, state secrets, etc. At the same time, it has great military significance in the fields of space remote sensing, navigation and positioning, and remote command operations [4–7]. The basic unit of digital image is pixel, and the means of image encryption is scrambling and diffusion of image pixels. Traditional encryption technology originates from the encryption of one-dimensional text data, and its encryption structure is designed for one-dimensional data flow

H. Chen · Y. Cao · Z. Yang (✉)
School of Space Information, Space Engineering University, Beijing 101416, China
e-mail: hangchen@alu.hit.edu.cn

H. Chen
e-mail: hitchenhang@foxmail.com

S. Liu · Z. Liu
Department of Automation Measurement and Control, Harbin Institute of Technology, Harbin 150001, China

H. Chen
Universitéde Lorraine, CNRS, CRAN UMR 7039, Nancy 54000, France

© The Author(s), under exclusive license to Springer Nature Switzerland AG 2023
H. Chen and Z. Liu (eds.), *Recent Advanced in Image Security Technologies*,
Studies in Computational Intelligence 1079,
https://doi.org/10.1007/978-3-031-22809-4_4

[8, 9]. Therefore, when encrypting an image, the image needs to be converted to one-dimensional data flow first, and then converted to an image format, which increases the image processing time and leads to low encryption efficiency. In addition to traditional encryption technology, there are many methods to realize pixel scrambling and diffusion, among which chaotic transformation [10–18] and optical transformation [5, 19–30] are the main ones. Image encryption using chaotic system is generally divided into two steps of scrambling and diffusion, and the initial value sensitivity and unpredictability of chaos can play an important role in both steps. Therefore, many encryption schemes combining chaos and chaos have been proposed by experts and scholars. Optical image encryption technology has the advantages of parallel processing, large key space, strong robustness, fast computing speed and high design freedom. The most prominent advantages are parallel processing, large key space and fast computing speed.

The Arnold transform is the most common chaos transform, including two-dimensional (2D) and three-dimensional (3D) Arnold [31, 32]. But there are some problems in 2D Arnold transform-based gray scale and color image encryption, such as simple low-dimensional chaos structure, insufficient key space, short period, cascade chaos security depends on unilateral chaos, etc. In hyperspectral image encryption based on 3D Arnold transform, there are some problems, such as using the same key in different bands and the encryption object must be square. Aiming at the above difficulties, this paper proposes three optical image encryption algorithms based on the improved Arnold model combined with Gyrator transform in optical transformation.

The rest of the chapter is organized in the following sequence. In Sect. 1, the general introductions of three improved Arnold models are presented. The three kinds of optical image encryption algorithms based on improved Arnold map in Gyrator domains are summarized in Sect. 2. In Sect. 3, the concluding remarks are given.

2   Three Kinds of Improved Arnold Maps

Compared with other chaos models, the Arnold model has two major advantages: one is that it can directly act on the image to save the time of image preprocessing; the other is that it can be transformed into 3D Arnold. However, the Arnold model also has some shortcomings. In order to overcome these shortcomings and meet higher safety requirements, three improved Arnold models are proposed in this paper.

2.1 Modified Arnold Mapping Model with Variable Parameters (M-Arnold)

Arnold transform is the most commonly used two-dimensional nonlinear scrambling system. The traditional two-dimensional Arnold diagram is mathematically defined as (1). As with other chaotic models, small changes in equation parameters and initial values may lead to large differences in qualitative structure of orbits. However, the parameters in the traditional Arnold model are fixed, which leads to the small space of transformation key and vulnerability. And after the multi—wheel dislocation, periodic phenomenon will appear.

$$\begin{pmatrix} x_{n+1} \\ y_{n+1} \end{pmatrix} = \begin{bmatrix} 1 & p \\ q & pq+1 \end{bmatrix} \begin{bmatrix} x_n \\ y_n \end{bmatrix} (\mathrm{mod}\, L) \tag{1}$$

where p and q are control parameters of the chaotic system, and they are settled once assigned. The system is the classic Arnold transform when $p = q = 1$ and transformation is equivalent to cutting and splicing of images, as shown in Fig. 1. (x_n, y_n) and (x_{n+1}, y_{n+1}) are the position coordinates of pixels before and after the transformation.

In order to overcome the problem of short cycle and small key space, the Arnold model is modified in this paper, and the formula is as follows

$$\begin{pmatrix} x_{n+1} \\ y_{n+1} \end{pmatrix} = \begin{bmatrix} 1 & x(p) \\ x(q) & x(p)*x(q)+1 \end{bmatrix} \begin{bmatrix} x_n \\ y_n \end{bmatrix} (\mathrm{mod}\, L) \tag{2}$$

where two pixels $x(p)$ and $x(q)$ are two values of pixels from the plaintext. The position of the two pixels is relatively fixed in the plaintext. But after scrambling each time of a single wheel, the encrypted object will change, and the pixel values at p and q will also change. Then $x(p)$ and $x(q)$ will also change. This leads to a rapid increase in key space and period, eventually achieving the effect of One-Time-Pad (OTP).

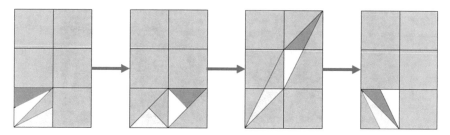

Fig. 1 The diagram of Arnold transformation

2.2 *LFSR-Cascaded Arnold Diffusion Model (LFSR-CADM)*

Arnold model can be used not just to scrambling, but also to diffusion. If diffusion is limited to Arnold, the security is fragile. In order to achieve higher safety requirements, this model cascades Arnold and Tent mapping to construct a new model. This model is called as cascaded Arnold diffusion model (CADM) and the mathematical expression is as follows

$$x_{n+1}(i) = x_n(i) + f(x_n(j)) + f(x_n(k)) \tag{3}$$

$$\begin{bmatrix} j \\ k \end{bmatrix} = \begin{bmatrix} 1 & p \\ q & pq+1 \end{bmatrix} \begin{bmatrix} i \\ i \end{bmatrix} (\mathrm{mod}\, L) \tag{4}$$

$$f(x_i) = \begin{cases} 2g_i + 1, & g_i \in [0, 2^{a-1}) \\ 2(2^a - 1 - g_i), & g_i \in [2^{a-1}, 2^a - 1] \end{cases} \tag{5}$$

$$g_i = (x_i + k_i) \bmod (2^a) \tag{6}$$

where, n is the number of iterative steps; $i = 1, 2, \cdots, L$ is grid point coordinates, L is system grid points. The boundary conditions of the model are determined by $x_n(0) = x_n(L)$ and $x_n(L+1) = x_n(1)$. And i, j, k are determined by the (4). a is the system digit, and k_i representing the horizontal movement distance of the mapping.

Cascade coupling greatly increases the complexity of key space and transformation, making diffusion difficult to crack. As shown in Fig. 2, (a) is transformed into a complete noise graph (b) by one time CADM.

To increase the overall security of the system, we associate chaos with traditional cryptography, and the linear feedback shift register is added to the system. The structure diagram of LFSR is shown in Fig. 3.

(a) the original image   (b) the encrypted image

Fig. 2 The scrambling effect of Arnold transform

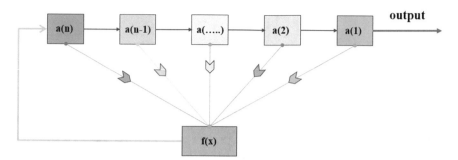

Fig. 3 The schematic diagram of a linear shift register

The pseudo-random sequence generated by LFSR controls the execution of different CADM, so that the whole system can generate multi-dimensional chaotic pseudo-random integer sequences rapidly and in parallel.

2.3 Block Mobile Model of 3D Arnold Map (BMM)

There are many kinds of images. In order to meet the requirements of image encryption in higher dimensions, scholars proposed three-dimensional Arnold on the basis of two-dimensional Arnold. The mathematical definition of a three-dimensional Arnold map can be written as (7). Based on the original model, this paper constructs the block mobile model of 3D Arnold map in (8).

$$
\begin{pmatrix} x_{n+1} \\ y_{n+1} \\ h_{n+1} \end{pmatrix} = \begin{bmatrix} 2 & 1 & 3 \\ 3 & 2 & 5 \\ 2 & 1 & 4 \end{bmatrix} \begin{bmatrix} x_n \\ y_n \\ h_n \end{bmatrix} (\mathrm{mod}\, L) \tag{7}
$$

$$
\begin{pmatrix} x_{n+1} \\ y_{n+1} \\ h_{n+1} \end{pmatrix} = \begin{bmatrix} 2 & 1 & 3 \\ 3 & 2 & 5 \\ 2 & 1 & 4 \end{bmatrix} \begin{bmatrix} x_n \\ y_n \\ h_n \end{bmatrix} (\mathrm{mod}\, h) \tag{8}
$$

where (x_n, y_n, h_n) and $(x_{n+1}, y_{n+1}, h_{n+1})$ are the position coordinates of pixels before and after the transformation. As shown in (7), L is size of the image; and in (8) h is the number of bands in the hyperspectral image.

Figure 4a and b show the process and result of the block mobile method. As the block steps, different areas will gradually be disorganized. The end result is that 3D Arnold performs different times in different areas of the image, so that a higher level of security is achieved. More importantly, this method is easy to implement and can be applied to the vast majority of hyperspectral images.

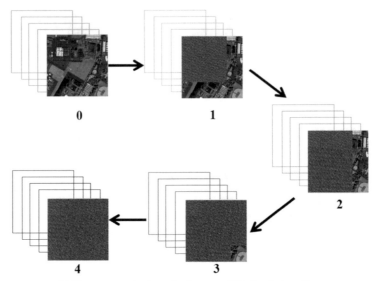

(a) The movement schematic diagram of 3D Arnold transform

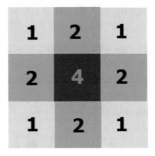

(b) The result of transformation and the number of transformations in different regions

Fig. 4 The block mobile method of 3D Arnold

2.4 Triangular Association Encryption Model (TAEM)

3D Arnold can realize the scrambling of 3D images, as well as the pixel diffusion of 3D images in space. In order to expand the pixel diffusion between the bands, a triangular association encryption model (TAEM) is constructed based on 3D Arnold. This model uses 3D Arnold to transform position three times in 3D space and find three points. Tent mapping is made for the pixel values of the three points, and then the three points are coupled. The mathematical expression of TAEM is shown below:

$$
\begin{aligned}
I[x_{n+1}(i), &\, y_{n+1}(i), h_{n+1}(i)] \\
&= g\{I[x_n(i), y_n(i), h_n(i)]\} + g\{I[x_n(i+a), y_n(i+a), h_n(i+a)]\} \\
&\quad + g\{I[x_n(i), y_n(i), N+1]\} \mod N
\end{aligned} \tag{9}
$$

$$\begin{pmatrix} x_n(i+a) \\ y_n(i+a) \\ h_n(i+a) \end{pmatrix} = \begin{bmatrix} 2 & 1 & 3 \\ 3 & 2 & 5 \\ 2 & 1 & 4 \end{bmatrix}^a \begin{bmatrix} x_n(i) \\ y_n(i) \\ h_n(i) \end{bmatrix} \pmod{N} \tag{10}$$

$$g(x_k) = \begin{cases} 2f_k + 1, & f_k \in [0, 2^{b-1}) \\ 2(2^b - 1 - f_k), & f_k \in [2^{b-1}, 2^b - 1] \end{cases} \tag{11}$$

$$f_k = (x_k + j_k) \bmod (2^b) \tag{12}$$

where, $[x_{n+1}(i), y_{n+1}(i), h_{n+1}(i)]$ is the pixel coordinates after transformation, and $I[x_{n+1}(i), y_{n+1}(i), h_{n+1}(i)]$ is the corresponding pixel value; $[x_n(i), y_n(i), h_n(i)]$ is the pixel coordinates before transformation, and $I[x_n(i), y_n(i), h_n(i)]$ is the corresponding pixel value; $[x_n(i), y_n(i), N+1]$ is the pixel coordinates at layer $N + 1$ corresponding to $[x_n(i), y_n(i), h_n(i)]$, and $I[x_n(i), y_n(i), N+1]$ is the corresponding pixel value; n is the number of iterative steps; $i = 1, 2, \cdots, N$ is number of transformation; N is system grid points. The three points in space are shown in Fig. 5. And $[x_n(i + a), y_n(i + a), h_n(i + a)]$ are determined by the model (8). In (9) and (10), b is the system digit, and j_k representing the horizontal movement distance of the mapping.

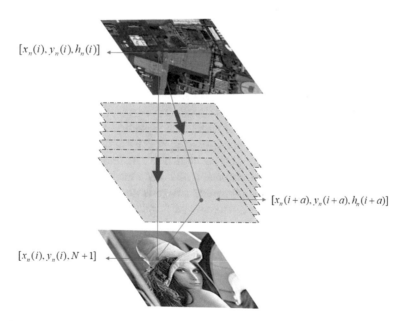

$[x_n(i), y_n(i), h_n(i)]$

$[x_n(i+a), y_n(i+a), h_n(i+a)]$

$[x_n(i), y_n(i), N+1]$

Fig. 5 A triangle in space formed by three points of the image from the algorithm (TAEM)

3 Optical Cryptosystem Based on Arnold Map and Improved Transformations in Gyrator Domains

The three algorithms proposed in this article are all implemented in the Gyrator domain, so Gyrator will be briefly introduced next. Gyrator transform is not only a generalized Fourier transform, but also a special linear regular transform. The Gyrator transform is mathematically defined as:

$$
\begin{aligned}
G(u, v) &= \xi^\alpha[g(x, y)](u, v) \\
&= \frac{1}{|\sin\theta|} \iint g(x, y) \exp\left[i2\pi \frac{(xy + uv)\cos\theta - xv - yu}{\sin\theta} \right] dx dy
\end{aligned}
\tag{13}
$$

where $g(x, y)$ and $G(u, v)$ represent the original image and the result image. (x, y) and (u, v) are the input spatial position coordinate and the frequency coordinate of the transformation domain. Besides the parameter α represents the fractional order of Gyrator transformation, that is, the rotation angle.

In addition, the computer environment in which the experiment is run is Windows10 system, Intel (R) Core (TM) i7-10,700 CPU @ 2.90 GHz, and 8.00 GB of RAM.

3.1 An Image Encryption and Hiding Algorithm Based on M-Arnold Model in Gyrator Domains

In this section, a gray image encryption and color image hiding algorithm based on M-Arnold model in Gyrator domains is introduced. Firstly, the original gray image will be scrambled by modified Arnold with variable parameters (M-Arnold). Subsequently, the obtained intermediate data are encrypted by cascaded Arnold diffusion model (CADM). The ciphertext image is hidden in bits into the RGB three bands of the color image. Then, the color image with hidden message is processed by M-Arnold in Gyrators and the final ciphertext image is obtained. The corresponding grayscale image hidden can be recovered by performing along the reverse direction of the encryption process. The validity and capability of the proposed encryption and hiding algorithm have been verified by numerical simulation results.

An improved Arnold model which can realize One-Time-Pad (OTP) is proposed. This section introduces cascaded Arnold diffusion model (CADM) and an idea which hides gray image ciphertext into color image in Gyrator domains. Firstly, the original gray image will be scrambled by M-Arnold which can implement One-Time-Pad (OTP). Subsequently, the obtained intermediate data are encrypted by CADM which can cascade 2D Arnold and tent mappings. M-Arnold and CADM greatly increase the key space of the algorithm and the period of Arnold model, enhancing the security of the algorithm. The image is divided into three groups of bits $(E_1)(E_2, E_3, E_4)(E_5, E_6, E_7, E_8)$, where (E_1) is the highest. Then three groups

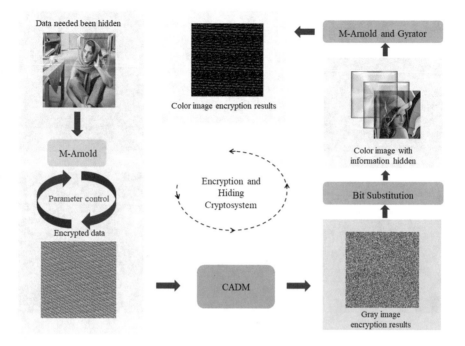

Fig. 6 The flowchart of gray image encryption and color image hiding system

are hidden into the RGB three bands of the color image. Finally the color image is processed by M-Arnold in Gyrators. After an attacker has obtained a ciphertext color image or decrypted color image, the hidden gray image cannot be seen. The flowchart of the proposed encryption scheme is illustrated in Fig. 6.

The proposed technique is tested by a series of simulations. The next experiment takes the gray image "Barbara" with 256*256 pixels as the original secret image, and the color image "Lena" with 256*256*3 pixels as the original hiding color image, shown in Fig. 7a and b.

From Fig. 7c and d, the original gray image can be successfully encrypted and the encrypted data can be hidden into the color image well. Besides the encrypted image is random pattern and there is no visual difference between the image containing the gray image and the original color image. And the encrypted image of Fig. 7d is shown in Fig. 7e. The gray image recovered from the color image and decrypted is shown in Fig. 7f, which is equal to the original one in human vision.

In the following step, we will test the security of the proposed hyperspectral image encryption scheme. Firstly, the key sensitivity test is performed. Both the parameters in M-Arnold transform, CADM and the rational angle in gyrator can be considered as the keys in these encrypted images, as mentioned above.

In simulation p, q, ki are regarded as the keys to decrypt the secret data and the result is depicted in Fig. 8. When only one of the keys is changed, the image will not

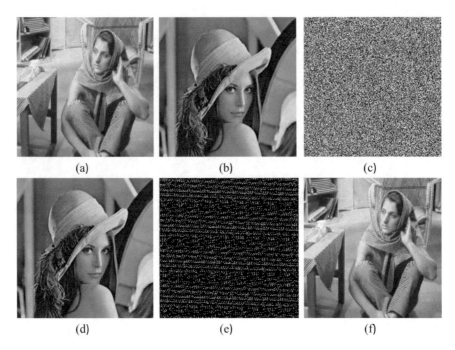

Fig. 7 The original, process and the final restored image: (**a**) the original gray image, (**b**) the original color image, (**c**) the encrypted image of (**a**), (**d**) the color image with hidden data of (**c**), (**e**) the encrypted image of (**d**), (**f**) the decrypted and restored image from (**e**)

be decrypted correctly. Even if the range of change is small, SSIM value will change greatly, as shown in Table.1.

The occlusion attack is performed on pixel blocks at different positions in cipher-text. As shown in Fig. 9a–d, when the encrypted color image or the color image with hidden information are occluded by 40*40 and 80*80, the plaintext information is

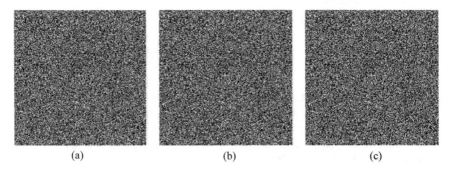

Fig. 8 The decrypted images with keys changed of MINCCM: (**a**) the decrypted image with key of p = 9, (**b**) the decrypted image with key of q = 11, (**c**) the decrypted image with key of ki = 4

Table 1 The results by using different initial conditions

p	q	ki	$SSIM$
10	10	5	1
9	10	5	0.0107
10	11	5	0.0113
10	10	4	0.1286

still clearly visible. This shows that the encryption algorithm can resist occlusion attack effectively.

The most important advantage of the encryption and hiding algorithms proposed in this paper is fraudulence. Even if an attacker decrypts a color image, it is difficult to realize that the color image contains an encrypted grayscale image. As shown in Fig. 10, the two images and their histograms are very alike visually. More important, the SSIM of different layers from the original color image and the color image with hidden gray image are 0.9968, 0.9438 and 0.8147.

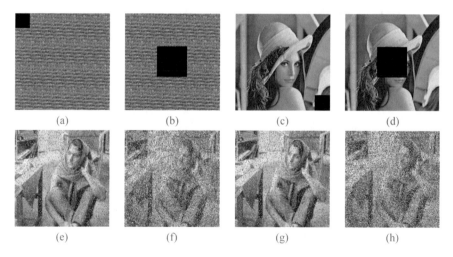

(a)   (b)   (c)   (d)

(e)   (f)   (g)   (h)

Fig. 9 The test of occlusion attack: (**a**) occluded part 40*40 of the encrypted color image, (**b**) occluded part 80*80 of the encrypted color image, (**c**) occluded part 40*40 of the color image with hidden data, (**d**) occluded part 80*80 of the color image with hidden data, (**e**) the gray image restored from (**a**), (**f**) the gray image restored from (**b**), (**g**) the gray image restored from (**c**), (**h**)the gray image restored from (**d**)

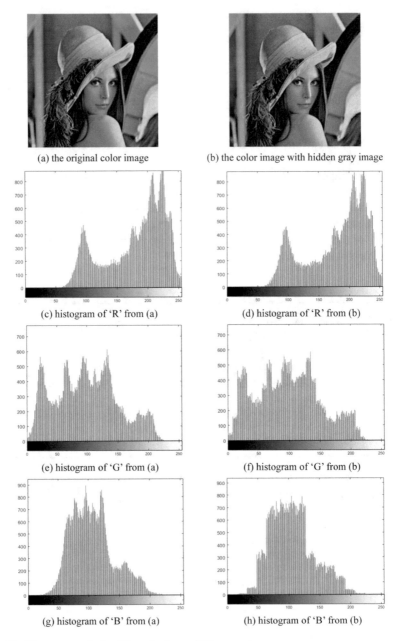

(a) the original color image   (b) the color image with hidden gray image

(c) histogram of 'R' from (a)   (d) histogram of 'R' from (b)

(e) histogram of 'G' from (a)   (f) histogram of 'G' from (b)

(g) histogram of 'B' from (a)   (h) histogram of 'B' from (b)

Fig. 10 Histogram analysis of the color image hiding gray image

3.2 Securing Color Image by Using Bit-Level LFSR-Cascaded Arnold Diffusion Model (LFSR-CADM) in Gyrator Domains

In this section, a color image cryptosystem is expressed based on bit-level LFSR-cascaded Arnold diffusion model (LFSR-CADM). In the encryption process, a pseudo-random sequence generated by LFSR is considered and utilized. In order to enhance the security of the encryption algorithm, two different bit scrambling schemes and different CADM models are set up. Then, the pseudo-random sequence of LFSR is used to control the selection of different encryption routes. Then, a Gyrator transform is performed on the encrypted image. In this cryptosystem, pseudo-random sequence is used as the main key, and Arnold, bit scrambling scheme, CADM, gyrator parameters are used as additional keys to enhance security. The performance of the proposed color image encryption algorithm is verified by numerical simulation. The flowchart of the proposed image encryption is depicted in Fig. 11.

In this scheme, the following points need to be paid attention to: LFSR uses the original polynomial to generate the pseudo-random sequence with the maximum period, and uses 01 bits in the pseudo-random sequence as the selector; for each selection, Arnold scramble is performed firstly; in bit-level scrambling scheme, different

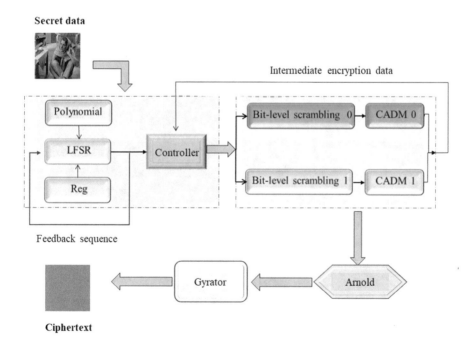

Fig. 11 The flowchart of LFSR-CADM color image encryption algorithm

adjustment of bit layer order determines different schemes; the difference between cadm models is Arnold's parameter setting.

In the following experiment, the color image 'Lena' with 256*256 pixels is used as the original secret color image, as shown in Fig. 12a. The encrypted image is shown in Fig. 12c, and the decrypted one is shown in Fig. 12d. From the encryption method described above, the original color image can be successfully encrypted and decrypted.

Firstly, we design some experiments are designed with incorrect keys to validate the sensitivity of the keys in protecting secret information. In this experiments, suppose that the complete encryption system (including encryption and decryption methods) and some of the keys are intercepted by illegal users.

Three key parameters are tested below, where the correct key parameters are as follows, reg = [1, 1, 0, 1, 0] in LFSR, whether to do bit-level scrambling is set as k = 1(otherwise k = 0), and q = 10 in CADM. The several key parameters above must be set to an integer, so the error key is also set to an integer. Subsequently, we

Fig. 12 Experiment data: (**a**) the original image, (**b**) intermediate encoded image, (**c**) encoded image, (**d**) the decrypted image

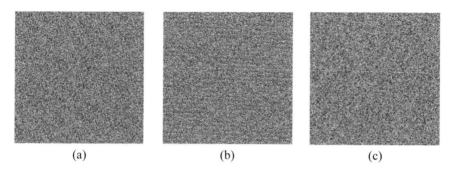

(a)   (b)   (c)

Fig. 13 Experiment data: (**a**) decrypted image with reg = [1, 1, 1, 1, 0], (**b**) decrypted image with k = 0, (**b**) decrypted image with q = 9

Table 2 The results by using different initial conditions

reg	k	q	SSIM
[1, 1, 0, 1, 0]	1	10	0.9998
[1, 1, 1, 1, 0]	1	10	0.1518
[1, 1, 0, 1, 0]	0	10	0.1553
[1, 1, 0, 1, 0]	1	9	0.1470

try to recover encrypted data by using the wrong key parameters and all the correct additional keys. The corresponding decryption graphs are shown in the Fig. 13, which all are meaningless chaos. And the SSIM between three decryption graphs and the original image are shown in Table 2.

The robustness analysis contains noise and occlusion attacks. First, a noise model is considered and introduced to complete the noise attack robustness experiments:

$$I'(x, y) = I(x, y)[1 + p \cdot \partial_{0,1}(x, y)] \tag{13}$$

where $I(x, y)$ and $I'(x, y)$ are the test images before and after adding noise respectively. The symbol p_1 indicates the noise intensity factor, and the larger symbol p_2 indicates higher noise intensity. In addition, $\partial_{0,1}(x, y)$ is random data with mean value of 0 and standard deviation of 1, which is the same size as the test image.

As shown in Fig. 14, we can plot a SSIM curve by decrypting the noise-added ciphertext generated with different intensity factor values. In the process, calculation uses coefficients from 0 to 0.1 with step size of 0.001. When the noise level is 0.001 and 0.002, we can recognize the main information of the images.

The next step will perform the occlusion attack experiment. In the simulation, we assume that the attacker know both the decryption scheme and the decryption key. This algorithm can resist shearing attack well, as shown in the Fig. 15.

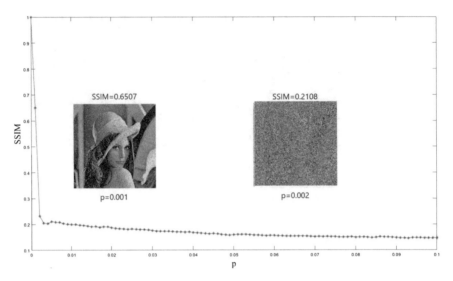

Fig. 14 The SSIM curve of noise attack including decrypted image obtained with p = 0.001, and decrypted image obtained with p = 0.002

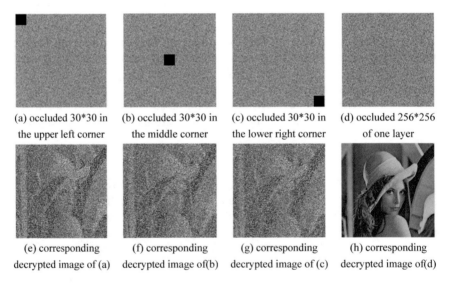

(a) occluded 30*30 in the upper left corner   (b) occluded 30*30 in the middle corner   (c) occluded 30*30 in the lower right corner   (d) occluded 256*256 of one layer

(e) corresponding decrypted image of (a)   (f) corresponding decrypted image of(b)   (g) corresponding decrypted image of (c)   (h) corresponding decrypted image of(d)

Fig. 15 The test of occlusion attack: (**a–d**) the occluded cipher text, (**e–h**) retrieved image

To verify the security of the cryptosystem, next experiments perform the known plaintext attack [33] and the chosen plaintext attack [34]. Because they are the most widely used schemes among the existing attack schemes. First of all, an encryption model is defined and expressed as follows

$$EC(x, y) = FN^{\alpha}\{ \exp[i \cdot \delta_1(x, y)] * IM(x, y)\} \exp[i \cdot \delta_2(x', y')] \qquad (14)$$

where the symbol FN^{α} denotes the Fresnel transform with rotation angle α. Besides, the functions $\delta_1(x, y)$ and $\delta_2(x', y')$ represent two random phase masks. And the function $EC(x, y)$ is considered as the RGB components of the ciphertexts in this paper.

Accordingly, this experiment takes the iterative phase retrieval algorithm and impulse function as the known plaintext attack [33] and chosen plaintext attack [34], respectively. Figure 16a and b are the two test images 'Lena' and 'Barbara' having 128*128 pixels which will be encrypted by the proposed cryptosystem. Figure 16c and d are the encrypted results. And the attacker ends up with Fig. 16e from Fig. 16d, in the condition that Fig. 16a and c are known, by the known plaintext attack. Moreover, the attacker ends up with Fig. 16f from Fig. 16d, in the condition that Fig. 16a and c are known, by the chosen plaintext attack. Because Fig. 16e and f are in random mode, the algorithm can resist the known plaintext attack and the chosen plaintext attack.

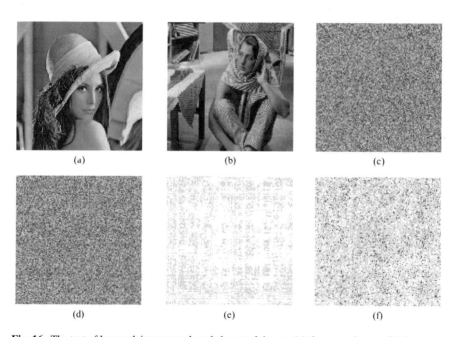

(a)   (b)   (c)

(d)   (e)   (f)

Fig. 16 The test of know plaintext attack and chosen plaintext: (**a**) the secret image, (**b**) the secret image, (**c**) the encrypted data of (**a**), (**d**) the encrypted data of (**b**), (**e**) the result of known plaintext attack and (**f**) the result of chosen plaintext attack

3.3 A Novel Signature and Authentication Cryptosystem for Hyperspectral Data by Using Triangular Association Encryption Algorithm in Gyrator Domains

This section presents an optical hyperspectral image cryptosystem based on triangular correlation encryption algorithm. This optical encryption scheme can hide spatial and spectral information simultaneously, that is, spatial scrambling and diffusion can be realized. Firstly, the signature identity image is appended to the original hyperspectral image. Then the hyperspectral image is spatially scrambled. Then, the triangular association encryption algorithm is performed on the Gyrator domain to generate a ciphertext graph with signature information. In this case, the signature image does not change. Finally, the signature image is encrypted by using the information of multiple points in the ciphertext image to obtain the ciphertext image of integrity verification information, which serves as the last layer of hyperspectral image.

The flowchart of the hyperspectral image encryption scheme in this chapter is illustrated in Fig. 17. The block mobile model of 3D Arnold map (BMM), triangular association encryption model (TAEM), Gyrator and the integrity authentication model are considered and utilized to complete this cryptosystem. Then make a brief introduction to the integrity verification algorithm and the mathematical formula is as (15). After signature encryption is performed, the authentication algorithm is performed separately for every layer. In this algorithm, if the original information is changed as little as one pixel, the SSIM value will also drop below 0.6(see numerical simulation).

$$
\begin{aligned}
I[x_{n+1}(i), \; & y_{n+1}(i), h_{n+1}(i)] \\
= g\{ & I[x_n(i), y_n(i), h_n(i)]\} \\
+ g\{ & I[x_n(i+1), y_n(i+1), h_n(i+1)]\} \\
+ g\{ & I[x_n(i+2), y_n(i+2), h_n(i+2)]\} \\
+ g\{ & I[x_n(i+3), y_n(i+3), h_n(i+3)]\} \\
+ g\{ & I[x_n(i+4), y_n(i+4), h_n(i+4)]\} \quad \text{mod } N
\end{aligned}
\tag{15}
$$

Then we carry out some numerical simulations to demonstrate the validity and capability of the proposed encryption algorithm. To complete the experiments, a hyperspectral image 'Sandiego' from AVIRIS having $256 \times 256 \times 189$ pixels is considered as the original image. And a grayscale image of the color image 'Lena' having $256 \times 256 \times 3$ pixels is considered as the signature image. The false RGB color composites consisting of the 30th, 70th and 100th band image and the color image are displayed in Fig. 18a and b, respectively. As shown in Fig. 18c, the encrypted image is a random pattern and the information cannot be recognized completely. In addition, the decrypted image is equal to the original image for human vision as shown in Fig. 18a.

The random parameters k and O in the block mobile model of 3D Arnold map (BMM), a in the triangular association encryptions and the rotation angle α in the

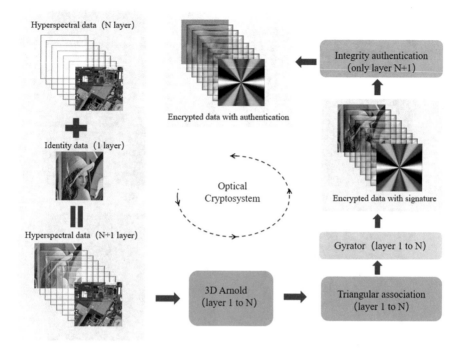

Fig. 17 The flowchart of the hyperspectral image encryption algorithm

gyrator transform can be regarded as the extra keys. Here, this experiment will analyze the contribution of the extra keys first. Some attackers are designed that have a complete encryption and decryption system, but with partial keys which are incorrect.

Note that all band images of the hyperspectral image are tested by the following experiment, however, for the sake of simplicity, the results shown here are only for false RGB color composites consisting of the 30th, 70th and 100th band image.

In calculation, the decryption process use the incorrect parameters and Fig. 18 shows the decrypted result. As we can see in Table 3, the correct keys are placed on the first line and the remaining lines have errors. As long as the key parameter is wrong, SSIM value will change greatly. As we can see from Fig. 19a–c), three noise-like decrypted images indicate that the tiny change in parameters k, a and O cause huge differences in the decrypted data. Therefore, the secret image will be protected by the extra keys perfectly.

Then next experiment will further tests the sensitivity of angle of the Gyrator transformation, and the corresponding SSIM curve between the original image and recovered image is depicted in Fig. 20a. In this test, the sampling step length of the additional key in experiment is taken at 0.005 and the decryption operation is performed 61 times with variable rotation angle α from 0.25 to 0.75. Suppose that the illegal user attacks the secret image by a random α, $\alpha=0.495$ and $\alpha=0.505$ for example, the decrypted result are displayed in Fig. 20a. Besides each band of

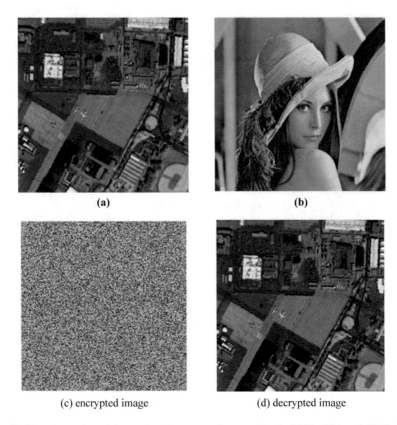

(a)   (b)

(c) encrypted image   (d) decrypted image

Fig. 18 Experiment data: (**a**) pseudo color composites combined of 30th, 70th and 100th band, (**b**) 'Lena' test image, (**c**) final encrypted image, (**d**) decrypted image

Table 3 The results by using different initial conditions

k	a	O	$SSIM - 30th$	$SSIM - 70th$	$SSIM - 100th$
3	3	1234	1	1	1
2	3	1234	0.0387	0.0371	0.0355
3	2	1234	0.0091	0.0109	0.0086
3	3	1324	0.0286	0.0308	0.0268

the decrypted image is displayed in Fig. 20b–g. Apparently, the decrypted result is blurred image and angle, as one of the additional keys, can protect the secret image well.

The robustness of the proposed method is verified by the noise attack and occlusion attack experiments. In the occlusion attack experiment, the decryption process is performed after the encrypted image occluded partly. And the calculation will fill the occluded section of the ciphertext with number 0. The ciphertext is occlude quarterly

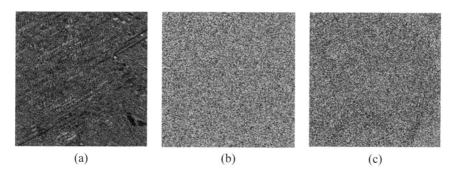

(a)   (b)   (c)

Fig. 19 Experiment data: (**a**) decrypted pseudo color composites combined of 30th, 70th and 100th band with k = 2, (**b**) decrypted ones with a = 2, (**c**) decrypted ones with O = 1324

and the corresponding result is shown in Fig. 21, in which the main information of the secret image can be recognized in vision.

Besides, the noise attack experiment is also done with the help of a noise adding model as (13). In the noise attack experiment, the SSIM value between noised added ciphertext and the original ciphertext is calculated. The SSIM curve generated with various value of coefficient from 0 to 1 with interval 0.01 is depicted in Fig. 22. Besides, the decrypted results with $p = 0.5$ and $p = 1$ are illustrated. Experimental results show that the cryptosystem can resist the noise attack and occlusion attack effectively.

In the following experiment, the encrypted algorithm will be tested by the known plaintext attack [33] and chosen plaintext attack [34]. To achieve the robustness analysis, a retrieval model is defined as (14). Accordingly, the iterative phase retrieval algorithm and impulse function can be used as the known plaintext attack and chosen plaintext attack, respectively.

Here the proposed cryptosystem encrypts the new layer-100 and layer-120 of 'sandiegou' having $256 \times 256 \times 60$ pixels displayed in Fig. 23a and b. The encrypted results of Fig. 23a and b are illustrated in Fig. 23c and d. It is assumed that the original image and its encrypted data are stolen by the attacker in this attack experiment. Then the simulation trys to attack the decrypted data of the layer-120 of 'sandiegou'. In calculation, each component in the known plaintext attack is iterated 500 times by phase retrieval algorithm. The extracted images given in Fig. 23e and f, which are obtained by known plaintext attack and chosen plaintext attack experiments, are noise-like and the secret information cannot be identified entirely.

In order to verify the validity of the signature algorithm, decryption is performed using the wrong signature image and the signature image of others. Figure 24a is the right signature image, Fig. 24b is other's signature image and Fig. 24c is the wrong signature image consisting of number 0. As shown in Fig. 24d–f, the image can only be decrypted correctly if the right signature image is used.

The validity of the integrity algorithm is verified below. We do experiments to simulate such an attacker who tries to tamper with the image information in the channel. Change different number of pixels in different positions of the image and

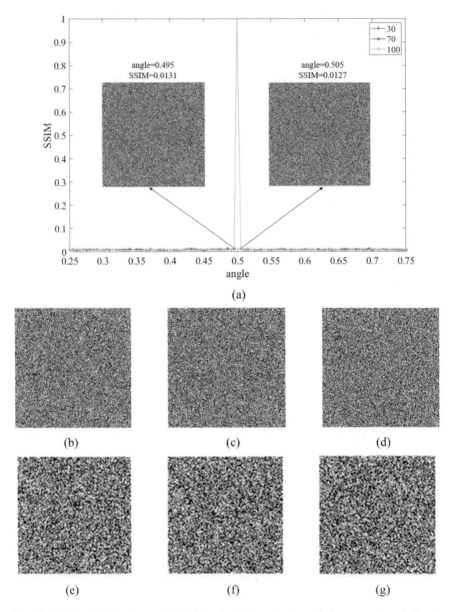

Fig. 20 (**a**) The SSIM curve calculated by the different values of the parameter of Gyrator (**b–d**) corresponding decrypted image of 30th, 70th and 100th band with angle = 0.495, (**e–g**) corresponding decrypted image of 30th, 70th and 100th band with angle = 0.505

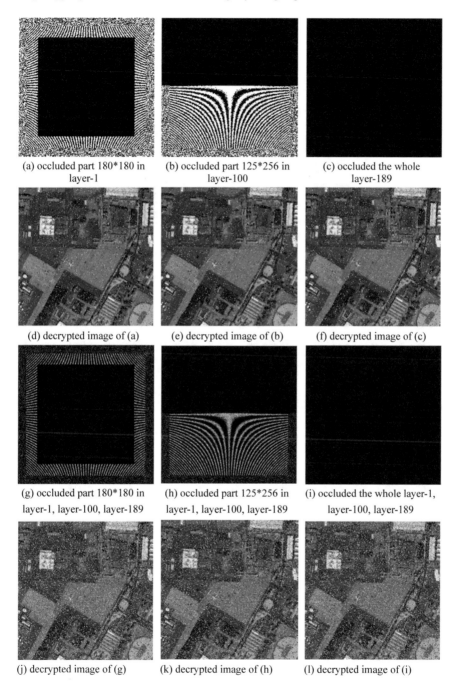

(a) occluded part 180*180 in layer-1

(b) occluded part 125*256 in layer-100

(c) occluded the whole layer-189

(d) decrypted image of (a)

(e) decrypted image of (b)

(f) decrypted image of (c)

(g) occluded part 180*180 in layer-1, layer-100, layer-189

(h) occluded part 125*256 in layer-1, layer-100, layer-189

(i) occluded the whole layer-1, layer-100, layer-189

(j) decrypted image of (g)

(k) decrypted image of (h)

(l) decrypted image of (i)

Fig. 21 The test of occlusion attack: (**a–c**) the occluded cipher text of layer-1, layer-100 and layer-189 respectively, and (**d–f**) corresponding decrypted image; (**g–i**) the occluded cipher text of the three layers, and (**j–l**) corresponding decrypted image

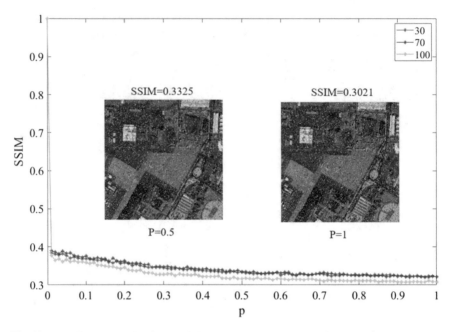

Fig. 22 The SSIM curve of noise attack including decrypted image obtained with $p = 0.5$, and decrypted image obtained with $p = 1$

Fig. 23 The test of know plaintext attack and chosen plaintext: (**a**) the secret image, (**b**) the secret image, (**c**) the encrypted data of (**a**), (**d**) the encrypted data of (**b**), (**e**) the result of known plaintext attack and (**f**) the result of chosen plaintext attack

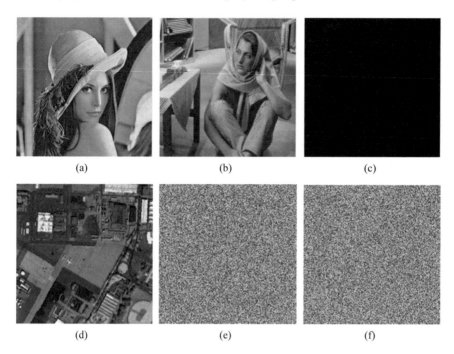

Fig. 24 (**a**) Image signature of 'Lena', (**b**) image signature of 'Barbara', (**c**) image signature of all zero, (**d–f**) the corresponding encrypted data with signature (**a**)–(**c**)

observe the changes of SSIM values between the modified image and the original image by using the integrity algorithm. As shown in Table 4, this algorithm is very sensitive to tampering, even if one pixel value is changed, SSIM value changes greatly. SSIM value decreases to blow 0.08 when nine elements are changed. It shows that the algorithm can resist tamper attack effectively.

4 General Conclusions

This chapter introduces three optical cryptography systems based on the improved Arnold and Gyrator systems. The effectiveness and performance of the three systems are verified by various numerical simulations. The experimental results show that the security of the cryptography system is improved by improving the randomness of Arnold system in Gyrator domain. Finally, it is verified that the three systems can be applied to gray scale, color and hyperspectrum respectively.

Table 4 The results with different numbers of pixels changed

x	y	h	$pixels`number - changed$	$SSIM - authentication$
0	0	0	0	1
75:75	75:75	99	1	0.5835
75:76	75:76	99	4	0.5276
75:77	75:77	99	9	0.0585
100:100	100:100	99	1	0.5934
100:101	100:101	99	4	0.1045
100:102	100:102	99	9	0.0706
75:75	75:75	189	1	0.5508
75:76	75:76	189	4	0.0764
75:77	75:77	189	9	0.0495
100:100	100:100	189	1	0.5680
100:101	100:101	189	4	0.0854
100:102	100:102	189	9	0.0719

References

1. Zhang C, Li H, Lu H (2020) Research on information encryption and hiding technology of 3D point cloud data model. In: 2020 International conference on computer science and management technology (ICCSMT)
2. Tunçer S (2016) Information encryption and hiding into an image by steganography methods to improve data security
3. Li J, Li J, Shen L (2014) Optical image encryption and hiding based on a modified Mach-Zehnder interferometer. Opt Express 22(4):4849
4. Ye HS, Zhou NR, Gong LH (2020) Multi-image compression-encryption scheme based on quaternion discrete fractional Hartley transform and improved pixel adaptive diffusion. Sign Process 175:107652
5. Qu G, Meng X, Yin Y, Wu H, Yang X, Peng X, He W (2021) Optical color image encryption based on Hadamard single-pixel imaging and Arnold transformation. Opt Lasers Eng 137:106392
6. Duan C, Zhou J, Gong , Wu J (2022) New color image encryption scheme based on multi-parameter fractional discrete Tchebyshev moments and nonlinear fractal permutation method. Opt Lasers Eng 150:106881
7. Liu W, Liu Z, Liu S (2013) Asymmetric cryptosystem using random binary phase modulation based on mixture retrieval type of Yang–Gu algorithm. Opt Lett 38(10):1651–1653
8. Hua Z, Zhu Z, Yi S, Zhang Z, Huang H (2021) Cross-plane color image encryption using a two-dimensional logistic tent modular map. Inf Sci 546:1063–1083
9. Liu X, Meng X, Wang Y, Yin Y, Yang X (2021) Known-plaintext cryptanalysis for a computational-ghost-imaging cryptosystem via the Pix2Pix generative adversarial network. Opt Express 29(26):43860–43874
10. Liu Z, Xu L, Liu T, Chen H, Li P, Lin C, Liu S (2011) Color image encryption by using Arnold transform and color-blend operation in discrete cosine transform domains. Optics Commun 284(1):123–128
11. Hamadi IA, Jamal RK, Mousa SK (2022) Image encryption based on computer generated hologram and Rossler chaotic system. Opt Quant Electron 54(1):1–12

12. Li M, Fan H, Xiang Y, Li Y, Zhang Y (2018) Cryptanalysis and improvement of a chaotic image encryption by first-order time-delay system. IEEE Multi-Media 25(3):92–101

13. Chen J, Han F, Qian W, Yao Y, Zhu Z (2018) Cryptanalysis and improvement in an image encryption scheme using combination of the 1D chaotic map. Nonlinear Dyn 93(4):2399–2413

14. Wang H, Xiao D, Chen X, Huang H (2020) Cryptanalysis and enhancement of an image encryption scheme based on a 1-D coupled Sine map. Nonlinear Dyn 100(1):1–15

15. Zhu C, Wang G, Sun K (2018) Improved cryptanalysis and enhancements of an image encryption scheme using combined 1D chaotic maps. Entropy 20(11):1–23

16. Zhang X, Wang X (2019) Multiple-image encryption algorithm based on DNA encoding and chaotic system. Multimed Tools Appl 78(6):7841–7869

17. Liu Z, Guo Q, Xu L, Muhammad Ashfaq Ahmad, Liu S (2010) Double image encryption by using iterative random binary encoding in gyrator domains. Opt Express 18(11):12033–43

18. Hu B, Guan Z, Xiong N, Chao H (2018) Intelligent impulsive synchronization of nonlinear interconnected neural networks for image protection. IEEE Trans Indus Inform 1–12

19. Alfalou A, Brosseau C (2009) Optical image compression and encryption methods. Adv Opt Photon 1(3):589–636

20. Liu S, Guo C, Sheridan JT (2014) A review of optical image encryption techniques. Opt Laser Technol 57:327–342

21. Chen H, Liu Z, Tanougast C, Liu F, Blondel W (2021) A novel chaos based optical cryptosystem for multiple images using DNA-blend and gyrator transform. Opt Lasers Eng 106448

22. Zhou J, Zhou N, Gong L (2020) Fast color image encryption scheme based on 3D orthogonal Latin squares and matching matrix. Opt Laser Technol 131:106437

23. Rakheja P, Singh P, Vig R, Kumar R (2020) Double image encryption scheme for iris template protection using 3D Lorenz system and modified equal modulus decomposition in hybrid transform domain. J Mod Opt 67(7):592–605

24. Yang X, Wu H, Yin Y et al (2020) Multiple-image encryption base on compressed coded aperture imaging. Opt Lasers Eng 127:105976

25. Duan CF, Zhou J, Gong L, Wu J, Zhou N (2021) New color image encryption scheme based on multi-parameter fractional discrete Tchebyshev moments and nonlinear fractal permutation method. Opt Lasers Eng 150:106881

26. Liu Z, Xu L, Lin C, Dai J, Liu S (2011) Image encryption scheme by using iterative random phase encoding in gyrator transform domains. Opt Lasers Eng 49(4):542–546

27. Li H, Bai X, Shan M, Zhong Z, Liu L, Liu B (2020) Optical encryption of hyperspectral images using improved binary tree structure and phase-truncated discrete multiple-parameter fractional Fourier transform. J Opt 22(5):055701

28. Chen H, Liu Z, Tanougast C, Blondel W (2021) Optical cryptosystem scheme for hyperspectral image based on random spiral transform in gyrator domains. Opt Lasers Eng 137:106375

29. Situ G, Zhang J (2004) Double random-phase encoding in the Fresnel domain. Opt Lett 29(14):1584–1586

30. Liu Z, Li S, Liu W, Liu W, Liu S (2013) Image hiding scheme by use of rotating squared sub-image in the gyrator transform domains. Opt Laser Technol 45(1):198–203

31. Wu C (2014) An improved discrete Arnold transform and its application in image scrambling and encryption. Acta Phys Sin 63(09):91–110

32. Wei K, Wen W, Fang Y (2020) Light field image encryption based on spatial-angular characteristic. Sign Process 2021(185):108080

33. Peng X, Zhang P, Wei H, Yu B (2006) Known-plaintext attack on optical encryption based on double random phase keys. Opt Lett 31:1044–1046

34. Peng X, Wei H, Zhang P (2006) Chosen-plaintext attack on lensless double-random phase encoding in the Fresnel domain. Opt Lett 31:3261–3263

Image Encryption Using a Chaotic/Hyperchaotic Multidimensional Discrete System

Camel Tanougast, Belqassim Bouteghrine, Said Sadoudi, and Hang Chen

Abstract With the evolution of the communication technology, fast and efficient tools for secure exchanged data are highly required in particular for image security. This chapter introduces a simplified and fast chaos-based scheme for multimedia data encryption and in particular for image security by a cipher application. The encryption algorithm is based on an extracted N-dimension (N-D) discrete time map. This considered chaotic/hyperchaotic discrete system includes several nonlinear terms and several parameters to generate a robust chaotic random key sequences satisfying image encryption requirements. The performance of the presented image encryption algorithm are analyzed by considering four important factors which are key space, correlation, complexity and running time. Results of the security analysis compared to similar proposals show that the described encryption scheme is more effective in terms of key stream cipher space, correlation, complexity and running time.

Keywords Chaos · Non-linear system · Multimedia · Encryption · Key space · Correlation · Running time · Image security

1 Introduction

Nowadays, image security is mainly achieved by the cryptography which most used for secure transmission of data. Indeed, image encryption is the most important issue for multimedia data transferred over insecure channel. The main goals that must be achieved during the transmission of information over the network is security by encrypting images with suitable encryption algorithms. An image encryp-

C. Tanougast (✉) · B. Bouteghrine
Lcoms, Université de Lorraine, Metz 57070, France
e-mail: camel.tanougast@univ-lorraine.fr

S. Sadoudi
Laboratoire Télécommunications, Ecole Militaire Polytechnique, Algiers 16111, Algeria

H. Chen
School of Space Information, Space Engineering University, Beijing 101416, China

tion converts a confidential image to a very difficult understanding and recognizing image that the content cannot to be obtained without a key for decryption. The security of image can be achieved by various types of encryption schemes that can be have some security and performance issues. Different chaos or non-chaos based encryption algorithms have been proposed. Usually, some non-chaos encryption schemes are standardized for some applications. However, the evolution of technologies and computing and memory resources constraints have made weak standardized encryption algorithms such as DES (Data-Encryption-Standard), 3DES (Triple Data-Encryption-Standard), IDEA (International-Data- Encryption-Algorithm) and EC (Elliptic-Curve) while lacking of robustness as security drawbacks (encryption key size, key exchange procedure) [1–5].

In recent decade, the chaotic based encryptions are considered to be more promising. The chaotic/hyperchaotic image encryptions are developed by exploit properties of deterministic dynamics and unpredictable behaviors of chaotic non-linear systems. Therefore, several chaotic encryption algorithms have been proposed for the encryption of data and images using known chaotic systems such as the systems of Lorenz [6], modified Arnold-Cat [7], Hénon [8–10], logistics [11], the Baker system [12] and the Ikeda system [13]. Chaotic systems can also be used to improve the strength of traditional AES (Advanced Encryption Standard) ciphers. For example, the latter has been combined with the chaotic systems of Hénon and Chebyshev to propose a new image encryption algorithm [14]. Similarly, chaotic systems are combined with the DNA encryption algorithm to increase its robustness and secure the transmission and storage of images [14, 15]. The basic principle of these combined algorithms is to convert the data to be encrypted into a DNA sequence, then to use a chaotic system to encrypt this data. The main drawback of these combined solutions is their long execution times limiting their use for real-time applications [18]. To reduce the computation time of the encryption process, these algorithms have been simplified by using discrete-time chaotic systems (such as Arnold-Cat, logistic and sinusoidal) and ensuring the performance achieved in terms of robustness and running time [19, 20]. However, the encrypted images obtained with these algorithms show a close correlation with the encryption key. Indeed, the chaotic Arnold-Cat, Hénon, sinusoidal, logistic and Chebyshev systems have been shown to be less suitable in combination for the key generation due to their limited distribution of output sequences [21]. In addition, these previous encryption algorithms have robustness weaknesses in terms of small space of generated keys, sensitivity, flaws against known plaintext and known ciphertext attacks, etc. [22].

This chapter presents an image encryption algorithm applying diffusion and confusion processes based on a multidimensional chaotic discrete system. The dynamic behaviors of the proposed chaotic map is investigated using trajectories graphs proving its suitability for image security. The described image encryption provides a better trade-off between the image security robustness and the ciphering scheme complexity.

The rest of this chapter is organized as follows. Section 2 describes the properties of the used multidimensional chaotic discrete system in the image encryption scheme. Section 3 details the chaos based encryption algorithm for the colour image security.

Section 4 gives the obtained performance of the proposed encryption solution proving its suitability for image encryption applications by considering the key space analysis and its resistance against some attacks. Finally, conclusion and future work are given in Sect. 5.

2 Multidimensional Chaotic Discrete System

To perform the image security by chaos encryption scheme, a novel multidimensional (MD) discrete time chaotic model defined by $2M - 1$ non-linear terms ($M > 2$) and M bifurcation parameters (controllers) is described by the follows recurrent equations [16]:

$$X_i(n+1) = 1 - a_i * X_i(n)^2 + \prod_{j=1; j \neq i}^{M} X_j(n); i < M$$

$$X_M(n+1) = a_M * \prod_{j=1}^{M-1} X_j(n)$$

(1)

Where X_i (for all $i=1...M$) are the state variables and a_i (for i=1..M) are the bifurcation parameters. Unlike all the proposed models of multidimensional chaotic systems presented in [16] and [18], the introduced model consists of M nonlinear subsystems containing $[2 * (M - 1) + 1 = 2M - 1]$ nonlinear terms and M control parameters. To design this model, the main property of the chaotic systems corresponding to the existence of a bounded solution is considered. Thereby, analyzing the system (1) in term of trajectories, the following points are figured out:

(i) By considering $X_i \in [-1,+1]$; as a result $\prod_{j=1}^{M-1} X_j(n) \in [-1,+1]$ implying the trajectory of the coordinate X_M is bounded for all values of the parameter a_M.
(ii) For all values of a_i, and $X_i \in [-1,+1]$, $1 - a_i * X_i$ is bounded; hence all solutions X_i are also bounded.

The analysis of the MD discrete system in terms of chaotic properties (Bifurcation, Lyapunov exponents, Trajectory, Kolmogorov entropy) is performed for dimensions 3, 4 and 5 [16]. Results prove that these 3-D, 4-D and 5-D systems generate more robust chaotic properties required for the encryption application purpose [17].

2.1 Mathematical Description of the 4-D Map

The 4-D discrete time chaos system having seven (07) nonlinear terms and four (04) control parameters is defined as follows [18]:

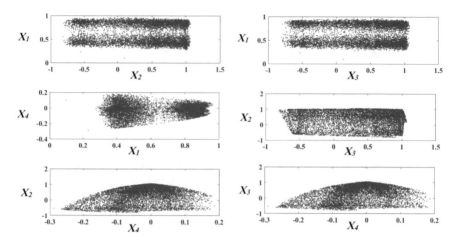

Fig. 1 Phase plan projections of the 4-D chaotic discrete system

$$X_1(n+1) = 1 - a_1 * X_1(n)^2 + [X_2(n) * X_3(n) * X_4(n)]$$
$$X_2(n+1) = 1 - a_2 * X_2(n)^2 + [X_1(n) * X_3(n) * X_4(n)]$$
$$X_3(n+1) = 1 - a_3 * X_3(n)^2 + [X_1(n) * X_2(n) * X_4(n)]$$
$$X_4(n+1) = a_4 * [X_1(n) * X_2(n) * X_3(n)]$$

$$(2)$$

Whereas X_1, X_2, X_3, X_4 are the state variables, $a_1, a_2, a_3, a_4 \in R$ are the controllers, and the integer n refers to the state.

2.2 Dynamical Behaviors of the 4-D System

To study dynamical behaviours of the 4-D system, the phase space trajectories as defined by the state variables (X_1, X_2, X_3, X_4) is used as an indicator to determine chaotic motions of the system [18]. Figure 1 shows the chaotic behaviour of the 4-D system obtained from the following values $a_1 = 1.6$; $a_2 = 0.78$; $a_3 = 1.45$ and $a_4 = -0.25$.

3 Algorithm for Colour Image Encryption

To achieve a better trade-off between security level and running time, a new two-stage chaos-based algorithm for colour image encryption is proposed. Unlike previous algorithms [17, 18], the considered image encryption is an one round algorithm including confusion and diffusion processes (see Fig. 2). Moreover, the image encryption algorithm is based on the extracted 4-D chaotic map to generate random keys stream cipher ensuring permutation and substitution processes. The main steps of the algorithm are summarized as follows:

Fig. 2 Flow chart of the
image encryption/decryption
algorithm

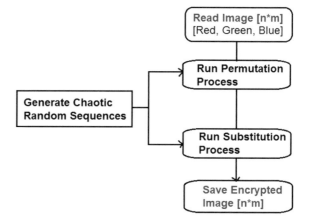

 (i) Read the input colour image and extract its size and pixel values of each color
     channel (Red, Green and Blue) in three matrices of values;
 (ii) Generate chaotic sequences used for substitution and permutation processes ;
(iii) For each color channel, run permutation and substitution algorithms;
(iv) The obtained matrices constitute the encrypted image.

The decryption algorithm is run in reverse way of the encryption algorithm.

Therefore, for an $n \times m$ input colour image, permutation and substitution algorithms are described as follows, respectively:

```
Read  Image(I(n,m));
Read  Image_Channels(Red,  Green,  Blue);
for  i:=1  to  n do
for  j:=1  to  m do
begin
{
Get_Permuted_Cell_Indices(d1,d2);
if  (Permutation_Condition(i,j)=='True'
   && Permutation_Condition(d1,d2)=='True') then
  {
  Permute_Cells  (Red(i,j),  Red(d1,d2));
  Permute_Cells  (Green(i,j),  Green(d1,d2));
  Permute_Cells  (Blue(i,j),  Blue(d1,d2));
  Permutation_Condition(i,j)='False';
  Permutation_Condition(d1,d2)='False';
  }
}
end;
```

```
Read Generated_Keys(Key1(X1), Key2(X2), Key3(X3));
for i:=1 to n do
for j:=1 to m do
begin
{
  Xor_Bit (Red(i,j), Key1(j+n(i−1)));
  Xor_Bit (Green(i,j), Key2(j+n(i−1)));
  Xor_Bit (Blue(i,j), Key3(j+n(i−1)));
}
end;
```

To show the efficiency of the proposed algorithm in terms of security level and running rime, the encryption algorithm is applied to encrypt some well-know images used in the literature. Figures 3, 4 and 5 show the obtained encrypted images using three different images.

Fig. 3 Original, encrypted and decrypted sailboat image

Fig. 4 Original, encrypted and decrypted Lenna image

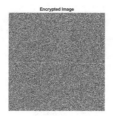

Fig. 5 Original, encrypted and decrypted Peppers image

4 Performance Analysis

To prove the efficiency while showing the obtained performance of the presented chaos-based algorithm, simulations and comparisons with previous works using similar environment and data set are presented. For the environment, the proposed algorithm is implemented using the well known programming language MATLAB (Version:R2017b). The developed platform is executed using an HP-Z400 workstation (3.2GHz processor and 8Go of RAM) under a Microsoft Windows 7 operating system. As a data set, some famous and similar images with different sizes are used to perform a fair comparison with previous works.

4.1 Key Space Analysis

Key space is considered as one of the most important metric used to prove the security of any encryption/decryption schemes. This metric indicates either the brute force attack fails or successes. In the proposed algorithm, the generated stream cipher key is based on 4-D chaos system defined by four (04) state variables and four (04) control parameters. Therefore, to generate a ten-digit floating sequence, the 4-D system covers a space of $10^{80} \approx 2^{300}$ satisfying the resistance condition of the well known exhaustive key search attack [15]. Consequently, as shown in Table 1, the extracted 4-D chaos map offers a more spacious key cipher value compared to some schemes introduced in similar research proposals.

4.2 Histograms Analysis

In any image, values of pixels are used by attackers to perform well known statistical attacks. To demonstrate the resistance against these attacks of the proposed algorithm, one comparison is given between the original image and its corresponding encrypted image using different histograms showing distributions of theirs values of pixels. Results shown in Figs. 6, 7 and 8 illustrate different distributions of values of pixels

Table 1 Key space comparison

Proposals	Key space value
[17]	2^{192}
[18]	2^{256}
[19]	2^{224}
[20]	2^{165}
[21]	2^{265}
[22]	2^{138}
[23]	2^{200}
[24]	2^{186}
[25]	2^{128}
[26]	2^{106}
Proposed	2^{300}

in original and corresponding encrypted images. Thereby, the uniform distribution observed in all channels of the encrypted image proves that the proposed algorithm resists to statistical attacks.

4.3 Correlation

In cryptography domain, image correlation metric is a method used for measuring the difference between the original image and its corresponding encrypted image relative to a given encryption scheme. Moreover, more this difference is very high, more the security level of the used cipher scheme is high. To analyse this metric, steps described in [17] are used. As shown in Figs. 9, 10 and 11, the propagation of two adjacent pixels in the Red, Green and Blue channels of the input image are not random as for the ciphered image. Consequently, these results indicate that pixels in the obtained cipher image are less correlated compared to the original picture. As a conclusion, the alliance observed among adjacent pixels in the ciphered image are hugely reduced compared to corresponding pixels in the original image. Numerically, three correlation indices are computed to evaluate the difference between the ciphered image and its corresponding original image. Tables 2, 3 and 4 synthesise obtained results and prove that even with a simplified and less complex scheme, the presented ciphering algorithm achieves lower correlations in the procured ciphered images.

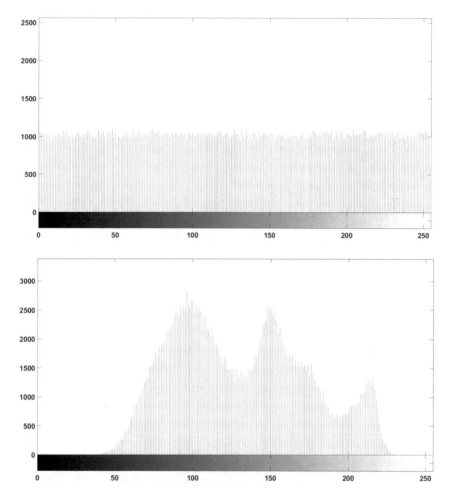

Fig. 6 Histogram comparison of the Red channel in Sail Boat image

4.4 Resistance to the Differential Attack

The rate of the number of pixels changing (or Number of pixels changing rate—*NPCR*) and the average intensity difference between the original and encrypted images (Unified average changing intensity—*UACI*) are parameters to evaluate the diffusion property of the image. These criteria evaluate the rate of change in the encrypted image by considering the change of a single pixel in the original image. To compute these parameters for an $n \times m$ image, the following equations are used:

$$NPCR = \frac{\sum_{i=1,j=1}^{n,m} D(i,j)}{(n \times m)} \times 100 \qquad (3)$$

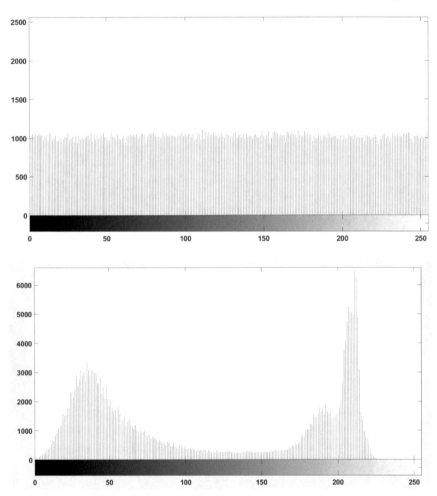

Fig. 7 Histogram comparison of the Blue channel in Sail Boat image

$$UACI = \frac{\sum_{i=1,j=1}^{n,m} \frac{|C1(i,j)-C2(i,j)|}{255}}{(n \times m)} \times 100 \qquad (4)$$

Where $C1$ and $C2$ are the $n \times m$ encrypted images before and after one pixel is changed in the plain image, respectively. As shown in Tables 5 and 6, the proposed image encryption scheme achieves a good scores for NPCR and UACI with only one round compared to previous similar works [27]. The obtained NPCR and UACI results are very close to the ideal score of random data with following values 99.609% and 33.463%, respectively.

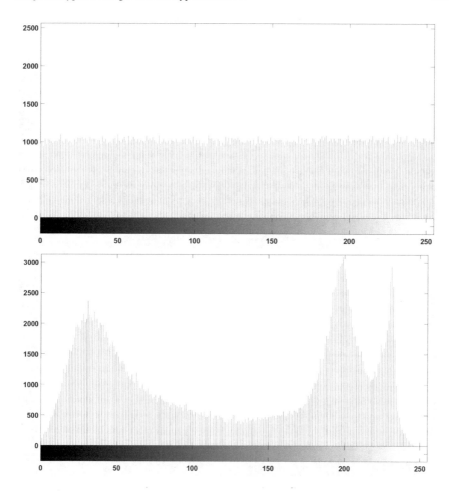

Fig. 8 Histogram comparison of the Blue channel in Sail Boat image

4.5 Cropping Attack Test

To evaluate the performance of the proposed image encryption algorithm against this kind of attack also known as data loss attack, different degrees of data loss tests to the decrypted image are performed in order to try to decrypt the cropped encrypted image. Results shown in Figs. 12, 13 and 14 prove that despite the encrypted image lost half of the main information, original image can be recovered. Therefore, the algorithm resist against data loss attack in different degrees.

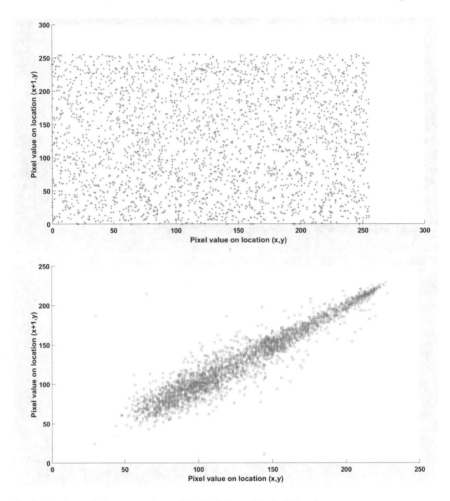

Fig. 9 Pixel correlation comparison of the Red channel in Sail Boat image

4.6 Entropy Information Analysis

Entropy information is the metric to evaluate the randomness of any data set. In image encryption domain, it measures the average information per bit in an image of $n \times m$ pixels [17]. Moreover, it is used to calculate the effectiveness of an image encryption scheme based on the following formula :

$$Entropy \; Value = - \sum_{i=1}^{n \times m} p(X_i) \log p(X_i) \qquad (5)$$

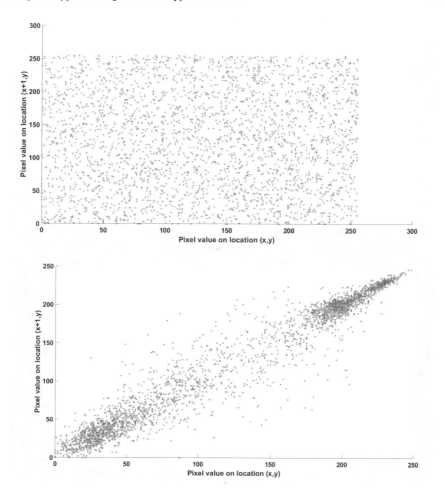

Fig. 10 Pixel correlation comparison of the Green channel in Sail Boat image

Where X_i is the value of the pixel at the position i, and $p(X_i)$ is its corresponding probability of appearance. Therefore, for a good encryption algorithm, computed entropy value of the encrypted image must be closer to 8. Results shown in Table 7 prove that the proposed scheme provides a better entropy value compared to previous similar works [8, 12, 15, 17, 27, 34, 39–41]. Moreover, the entropy value achieved with the proposed algorithm is very close to the ideal value.

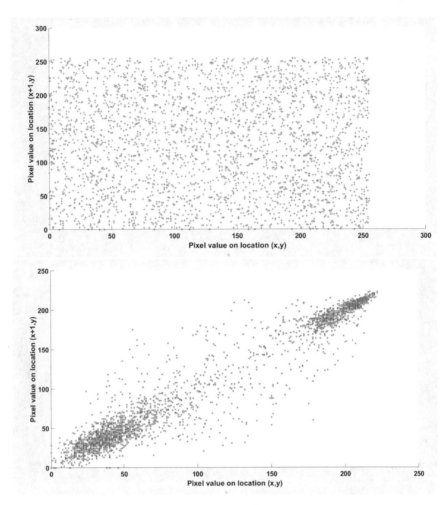

Fig. 11 Pixel correlation comparison of the Blue channel in Sail Boat image

Table 2 Correlation comparison between the original Sail boat image and its corresponding encrypted image

	Original			Encrypted		
	Red	Green	Blue	Red	Green	BLue
Diagonal	0.9415	0.9491	0.9571	0.0021	−0.0148	−0.0174
Vertical	0.9522	0.9706	0.9662	−0.0216	−0.0074	0.0096
Horizontal	0.9572	0.9696	0.9715	0.0142	0.0243	−0.0243

Table 3 Correlation comparison between the original Lenna image and its corresponding encrypted image

Type	Original			Encrypted		
	Red	Green	Blue	Red	Green	BLue
Diagonal	0.967	0.961	0.923	0.0144	−0.00013	0.0266
Vertical	0.9802	0.967	0.934	0.00657	−0.00973	0.00654
Horizontal	0.988	0.9803	0.958	−0.0003	0.0132	−0.0044

Table 4 Correlation comparison between the original Peppers image and its corresponding encrypted image

Type	Original			Encrypted		
	Red	Green	Blue	Red	Green	BLue
Diagonal	0.9539	0.9682	0.9547	−0.00278	0.0139	−0.00568
Vertical	0.9658	0.9817	0.9647	−0.00411	−0.022	−0.00604
Horizontal	0.9662	0.9843	0.9693	−0.0134	0.00069	−0.0191

Table 5 NPCR and UACI comparison

Proposals	NPCR (%)	UACI (%)
[19]	99,606	31,247
[20]	99,617	33,448
[24]	99,60	33,55
[27]	99,61	31,75
[33]	99,569	33,225
[40]	99,71	33,45
Proposed algorithm	99,605	33,307

Table 6 Number of round to achieve better NPCR and UACI comparison

Proposals	Permutations	Substitution
[27]	2	2
[28]	18	6
[29]	4	2
[30]	3	3
[31]	3	3
[18]	4	4
Proposed algorithm	1	1

Fig. 12 One eighth data loss attack test

Fig. 13 Quarter data loss attack test

Fig. 14 Half data loss attack test

4.7 Noise Resistance Test

To evaluate the performance of the proposed algorithm against several kinds of noise, several noises are incorporated to the encrypted image E. From the obtained decrypted image, the robustness of the proposed algorithm is measured. For the Speckle noise attack, a multiplicative noise is added by using the equation $E = E + r \times E$, where r is uniformly distributed random noise with mean equal to 0 and variance equal to 0.05. In case of salt and pepper noise, this test is performed with default noise density equal to 0.05 affecting approximately 5% of pixels in the encrypted image. Finally, a Gaussian noise test is applied by choosing a Gaussian white noise with a variance of 0.01. The results shown in Figs. 15, 16, 17 and 18 prove that the decrypted images still be recognized within a certain range of noises.

Table 7 Entropy information comparison

Proposals	Key space value
[24]	7.997
[25]	7.9973
[32]	7.99750
[33]	7.9024
[34]	7.99920
[27]	7.9972
[35]	7.9967
[36]	7.9973
Proposed algorithm	7.9992

Fig. 15 Obtained result after a Salt and Pepper noise attack test

Fig. 16 Obtained result after a Speckle noise attack test

Although the decrypted images are become fuzzier with the increase of noise intensity, the major information of the image is still figured out. Therefore, the proposed image encryption scheme resists against different kind of noise attacks.

4.8 Complexity

By considering only one round for the permutation and one round for the substitution, the proposed algorithm show less complexity than similar algorithms. To prove this performance, running speed comparison using the same data set (image input) as

Fig. 17 Obtained results after a Gaussian noise attack test

Fig. 18 Obtained results after a Poisson noise attack test

Table 8 Running time comparison

Proposals	256×256 image	512×512 image
[19]	0.823	3.253
[37]	1.256	4.828
[38]	1.205	4.750
[22]	/	3.64
[23]	/	22
[26]	/	1.82
[35]	0.464	1.708
[39]	/	3.42
[33]	0.538	1.908
[40]	/	4.02
[18]	0.387	1.644
Proposed algorithm	0.38	1.073

used in previous similar proposals is performed. According to the obtained results given in Table 8, the introduced chaos-based scheme for colour image encryption achieves better and faster running speed time than all the previous works [20, 23, 24, 34, 36, 38–41]. Therefore, this presented cipher scheme is more suitable for real time encryption applications.

5 Conclusion

This chapter detailed diffusion and confusion processes for the image encryption by using a deterministic ramdon system. More precisely, a chaotic/hyperchaotic multidimensional discrete system is used to generate ramdon sequences in algorithm applying diffusion and confusion processes for colour image encryption. The dynamic behaviors of the proposed chaotic map is investigated using trajectories graphs proving its suitability for image security. The described image encryption provides a better trade-off between the image security robustness and the ciphering scheme complexity. One main originality of the presented approach is to increase the level of robustness according to the order of the chaotic system and the encryption process. Indeed, as required for encryption purposes, the multidimensional discrete time chaotic system allows to produce more and more complex larger chaotic ranges in function of the dimension. A two-stage chaos-based algorithm is specifically designed for colour image ciphering. Simulation results prove the efficiency of one round scheme based on the chaotic discrete time system by satisfying encryption requirements and resisting to well known attacks such as the exhaustive key search, statistical and differential attacks. Moreover, the introduced image encryption provides high performance in term of execution running time suitable for real-time multimedia encryption applications compared to similar current schemes having low performance in term of security level.

References

1. Singh S, Sharma PK, Moon SY, Park JH (2017) Advanced lightweight encryption algorithms for IoT devices: survey, challenges and solutions. J Am Intell Human Comput 1–18
2. Salavi RR, Math MM, Kulkarni UP (2019) A survey of various cryptographic techniques: from traditional cryptography to fully homomorphic encryption. Innov Comp Sci Eng 295–305
3. Saraf KR, Jagtap VP, Mishra AK (2014) Text and image encryption decryption using advanced encryption standard. Int J Em Trends Technol Comp Sci (IJETTCS) 3:118–126
4. Mathur N, Bansode R (2016) AES based text encryption using 12 rounds with dynamic key selection. Proc Comp Sci 79:1036–1043
5. Singh LD, Singh KM (2015) Implementation of text encryption using elliptic curve cryptography. Proc Comp Sci 54:73–82
6. Singh U, Garg U (2013) An ASCII value based text data encryption system. Int J Sci Res Publ 3:2250–3153
7. Chandra S, Mandal B, Alam SS, Bhattacharyya S (2015) Content based double encryption algorithm using symmetric key cryptography. Proc Comp Sci 57:1228–1234
8. Iyer SC, Sedamkar RR, Gupta S (2016) A novel idea on multimedia encryption using hybrid crypto approach. Proc Comp Sci 79:293–298
9. Dixit P, Gupta AK, Trivedi MC, Yadav VK (2018) Traditional and hybrid encryption techniques: a survey. Netw Commun Data Knowl Eng 239–248
10. Al-Dahhan RR, Shi Q, Lee GM, Kifayat K (2019) Survey on revocation in ciphertext-policy attribute-based encryption. Sensors 19:1695
11. Kalmani VH, Goyal D, Singla S (2015) An efficient and secure solution for attribute revocation problem utilizing CP-ABE scheme in mobile cloud computing. Int J Comp Appl 129:16–21

12. Pang L, Yang J, Jiang Z (2014) A survey of research progress and development tendency of attribute-based encryption. Sci World J
13. Philip M, Das A (2011) Survey: image encryption using chaotic cryptography schemes. In: IJCA Special issue on computational science-new dimensions and perspectives, NCCSE, pp 1–4
14. Babaei M (2013) A novel text and image encryption method based on chaos theory and DNA computing. Nat Comp 12:101–107
15. Rachmawati D, Tarigan JT, Ginting ABC (2018) A comparative study of message digest 5 (MD5) and SHA256 algorithm. J Phys Conf Ser 978:012116
16. Alvarez G, Li S (2006) Some basic cryptographic requirements for chaos-based cryptosystems. Int J Bifurcation Chaos 16:2129–2151
17. Bouteghrine B, Tanougast C, Sadoudi S (2021) Design and FPGA implementation of new multidimensional chaotic map for secure communication. J Circ Syst Comp 2150280
18. Bouteghrine B, Tanougast C, Sadoudi S (2021) Novel image encryption algorithm based on new 3-d chaos map. Multimed Tools Appl 80:25583–25605
19. Bouteghrine B, Tanougast C, Sadoudi S (2021) Fast and efficient Chaos-based algorithm for multimedia data encryption. In: 2021 International conference on electrical, computer, communications and mechatronics engineering (ICECCME), IEEE, pp 1–5
20. Çavuşoğlu Ü, Kaçar S, Pehlivan I, Zengin A (2017) Secure image encryption algorithm design using a novel chaos based S-Box. Chaos, Solitons and Fractals 95:92–101
21. Wang X, Gao S, Yu L, Sun Y, Sun H (2019) Chaotic image encryption algorithm based on bit-combination scrambling in decimal system and dynamic diffusion. IEEE Access 7:103662–103677
22. Fu C, Zhang GY, Zhu M, Chen Z, Lei WM (2018) A new chaos-based color image encryption scheme with an efficient substitution keystream generation strategy. Secur Commun Netw
23. Pak C, Huang L (2017) A new color image encryption using combination of the 1D chaotic map. Sign Process 138:129–137
24. Liu H, Kadir A, Sun X (2017) Chaos-based fast colour image encryption scheme with true random number keys from environmental noise. IET Image Process 11:324–332
25. Tong XJ, Zhang M, Wang Z, Ma J (2016) A joint color image encryption and compression scheme based on hyper-chaotic system. Nonlinear Dyn 84:2333–2356
26. Niyat AY, Moattar MH, Torshiz MN (2017) Color image encryption based on hybrid hyper-chaotic system and cellular automata. Opt Lasers Eng 90:225–237
27. Li C, Luo G, Qin K, Li C (2017) An image encryption scheme based on chaotic tent map. Nonlinear Dyn 87:127–133
28. Wang Y, Wong KW, Liao X, Chen G (2011) A new chaos-based fast image encryption algorithm. Appl Soft Comput 11:514–522
29. Lian S, Sun J, Wang Z (2005) A block cipher based on a suitable use of the chaotic standard map. Chaos, Solitons and Fractals 26:117–129
30. Wong KW, Kwok BSH, Law WS (2008) A fast image encryption scheme based on chaotic standard map. Phys Lett A 372:2645–2652
31. Mao Y, Chen G, Lian S (2004) A novel fast image encryption scheme based on 3D chaotic baker maps. Int J Bifurcation chaos 14:3613–3624
32. Xiao D, Liao X, Wei P (2009) Analysis and improvement of a chaos-based image encryption algorithm. Chaos, Solitons and Fractals 40:2191–2199
33. Farajallah M, El Assad S, Deforges O (2016) Fast and secure chaos-based cryptosystem for images. Int J Bifurcation Chaos 26:1650021
34. Hua Z, Zhou Y, Pun CM, Chen CP (2015) 2D Sine logistic modulation map for image encryption. Inform Sci 297:80–94
35. Kanso A, Ghebleh M (2012) A novel image encryption algorithm based on a 3D chaotic map. Commun Nonlinear Sci Num Simul 17:2943–2959
36. Zhu C, Gan Z, Lu Y, Chai X (2019) An image encryption algorithm based on 3-d dna level permutation and substitution scheme. Multimed Tools Appl 1–32

37. Zhu H, Zhao C, Zhang X (2013) A novel image encryption-compression scheme using hyper-chaos and Chinese remainder theorem. Sign Process Image Commun 28:670–680
38. Wang X et al (2019) S-box based image encryption application using a chaotic system without equilibrium. Appl Sci 9:781
39. Zhang XP, Guo R, Chen HW, Zhao ZM, Wang JY (2018) Efficient image encryption scheme with synchronous substitution and diffusion based on double S-boxes. Chin Phys B 27:080701
40. Herbadji D, Belmeguenai A, Derouiche N, Liu H (2020) Colour image encryption scheme based on enhanced quadratic chaotic map. IET Image Process 14:40–52
41. Wu X, Zhu B, Hu Y, Ran Y (2017) A novel color image encryption scheme using rectangular transform-enhanced chaotic tent maps. IEEE Access 5:6429–6436

Video Cryptosystem Using Chaotic Systems

Said Sadoudi, Camel Tanougast, Belqassim Bouteghrine, Hang Chen, and Salah Mihoub

Abstract In recent years, due to the rapid technological evolution of the communication networks (internet, social networks, Cloud, mobile communication, IoT, etc.), information exchanged have become the digital privacy of users. The use of the word privacy explains the most valuable data (or information) for people. In this context, video data represents the most exchanged information between users given the ease offered by Information and Communication Technologies (ICT) namely Internet networks and mobile technologies (3G, 4G and 5G). Moreover, the interception of data communicated by a malicious third party constitutes a serious threat to the privacy of users by having access to confidential data. Therefore, information security for video data becomes a major challenge in current and future communications systems. However, a video stream presents a very large amount of data compared to the image or others types of data, in addition to a number of images per second having be respected according to the used standard. Consequently, securing video transmission using the compression is a challenge to satisfy the constraints required in real-time transmission. This chapter highlights the advantage of chaos-based cryptography in the field of securing the video stream through the proposal of a chaotic cryptosystem for video encryption transmitted in real time. The presented proposal combines data compression and encryption in a single process while using a robust chaotic synchronization technique. Finally, a thorough security analysis is performed to validate the presented video cryptosystem.

Keywords Information security · Video encryption · Chaotic system · Chaos-based cryptography · Compression and encryption

S. Sadoudi (✉) · S. Mihoub
Laboratoire Télécommunications, Ecole Militaire Polytechnique, Bordj El Bahri, Algiers, Algeria
e-mail: said.sadoudi@emp.mdn.dz

C. Tanougast · B. Bouteghrine
LCOMS Laboratory, Université de Lorraine, Nancy, France
e-mail: camel.tanougast@univ-lorraine.fr

H. Chen
School of Space Information, Space Engineering University, Beijing, China
e-mail: hitchenhang@foxmail.com

© The Author(s), under exclusive license to Springer Nature Switzerland AG 2023
H. Chen and Z. Liu (eds.), *Recent Advanced in Image Security Technologies*,
Studies in Computational Intelligence 1079,
https://doi.org/10.1007/978-3-031-22809-4_6

1   Introduction

Nowadays, many approaches exists to ensure the different security services including the confidentiality, the integrity and the authentication. However, to guarantee the robustness of these approaches is increasingly necessary to ensure no compromision by attacks from malefactors. For this purpose, several measures were used to test performance of a cryptosystem. In particular for video encryption cryptosystem, the mainly considered criteria ensuring to provide the first security only is the confidentiality. Moreover, peculiarities of video streams are very large amount of data compared to images or others data and required transmission rates expressed by the number of images per second according to used standards. Thereby, the transmission of the video is a challenge for researchers and engineers in the field of telecommunications always trying to overcome. Therefore, the compression of the video becomes necessary to satisfy the constraints required by the transmission of the video. Video compression techniques appeared in the 90s, and the most recent standard is H.265 [1] and the most current used standard is H.264 [2] where most video security works have been done around this standard also considered in this chapter. This chapter focuses on the study of the encryption performance of a video stream by using several performance criteria existing in the literature intended for the image. It also studies additional other criteria to evaluate the security of a video stream cryposystem. Simulations of these performance criteria is providing by considering a chaotic cryptosystem while showing its suitability for video encryption which must satisfy constraints of real-time video transmission. Finally, a comparative study with other reference works is given to demonstrate the contribution of chaotic cryptosystems in real-time video encryption in terms of security robustness.

2   State of the Art in Video Security Methods

The field of cryptography and communications security continues to grow. It affects several aspects of practical life such as the security of online traffic, the security of banking transactions, and the encryption of the video stream transmission usually integrated into video conferences and calls. Several previous works have been proposed for securing offline video or for streaming video (in real time). These different approaches can be classified into three main categories as summarized below.

2.1   Complete Security Algorithms

Encryption is one of techniques for securing a video stream existing. The current trend is towards selective encryption algorithms. These methods consist in performing a *xor* operation between all data video with the encryption keys after compression and encoding by using a stream cipher or a block cipher such as DES, IDEA, AES, etc. [3]. In terms of security, it is the safest because all data are encrypted. However, the huge amount of data and time cost reduce the encryption speed. The enhanced full encryption algorithms are the Video Encryption Algorithm (VEA) and the Chaotic Sequence Cipher (CSC). The VEA [4] divides the wafers into odd and even parts, then it encrypts the even part with DES and the other half is the result of the *xor* of the latter and the odd part. Therefore, this algorithm reduces the original complexity to half. The CSC is another algorithm doing a *xor* between data and chaotic sequences generated by a chaotic generator. The advantage of this algorithm compared to the previous one is that it is faster. Indeed, it does not introduce a block cipher costing in execution time.

There are two methods to perform full encryption corresponding either encrypting data before compression or after compression where each of them has its own drawbacks apart from the long computation time required for full encryption. If one take for example full encryption before compression:

1. Compression generally introduces irreversible losses and operations. Therefore, it's impossible to decipher after compression;
2. After encryption, data are random and compression usually exploiting data redundancy becomes inefficient.

If we consider full encryption after compression, there are also disadvantages like the previous case :

1. After encryption, data are random. Therefore, encryption after compression reduces the efficiency of the used compression algorithm;
2. The video becomes incompatible with the format of the used compression standard. Consequently, a video decoder fails to decode the encrypted video.

2.2 Selective Encryption Algorithms

Selective encryption is the trend of current research. It consists of partially encrypting video data. More precisely, this approach only encrypts sensitive data having an important value in a video stream. This technique is generally used in the compression phase to gain complexity and increase the execution speed of the algorithm. Selective encryption algorithms can be classified into three categories as detailed below:

2.2.1 MPEG Video Encryption Algorithm

This section presents some algorithms that appeared before H.264 standard.

Images Ciphering MPEG video stream security was proposed by [5]. The main idea of this algorithm is to encrypt the most important frames in the video which are I frames, because P and B frames are useless without know the corresponding I frames. However, Agi and Gong [6] have demonstrated that certain important parts of the video can be recovered independently of the I frames. Indeed, the cryptosystem does not take into account the I blocks contained in P and B frames. However, the compression and bit rates decrease significantly when the I-frame values are encrypted although one of the most important requirement of video compression is to increase the bit rate. To maintain the bit rate, the encrypted and original values must have the same length. Moreover, the encrypted value must not affect the length of other values. This requirement (constant bit rate) is more complex to be achieve if encryption is performed without taking entropy coding into consideration.

Four Levels of Security According to the work presented in [7], four levels of encryption algorithm for video have been implemented. The first one is to apply the encryption process on all headers. The second level is to encrypt the DC and AC coefficients of the I blocks after performing a DCT transformation. The third level is to encrypt the I, P and B frames. The fourth level is to encrypt all data. The implemented video encryption algorithm uses DES standards to encrypt MPEG-1. The main drawback of this algorithm is the problem of compatibility with the coder/decoder standard (i.e. it is not a format that complies with the standard used). Thus, a specific decoder is required to process the encrypted bitstream. In addition, the compression ratio decreases significantly, and the bit rate increases in this algorithm.

Other Permutation than the Zig-Zag Always around algorithms intended for MPEG-1, an algorithm based on a permutation of macroblocks (MB) are proposed. It consists of the following steps:

1. Form a vector of 64 coefficients where the first value is DC, and other 63 values are AC with the last value being zero [8].
2. The first bit of DC presents the LSB and the last bit represents the MSB.
3. A random permutation is performed for 64 values of the previous vector.

The disadvantage of this algorithm [9] is its vulnerability against known plain-text and known cipher-text attacks. Furthermore, it introduces compatibility problems and decreases the compression ratio while increasing the bit rate.

Algorithms based on DCT Coefficients Sign Change Tosun and Feng devised three levels of security for streaming MPEG-2 video by using 64 coefficients of the DCT [10]. This algorithm divides 64 values into three intervals: the first interval is $[0,...,x_1]$, the second one is $[x_1,..., x_2]$, and the last one is $[x_2,...,63]$. Depending on the level of security required, the first range or the first two ranges or all the ranges are chosen to encrypt. The proposed system encrypts coefficients before the entropy, but it decreases the compression ratio while increasing the execution time and the bit rate.

Other work on the frequency domain is carried out by Zeng and Lei [11]. They proposed a common selective encryption and compression (CSC) scheme for H.263. The proposed scheme inverts the sign of the coefficients, because by inverting the sign, the compression rate and the bit rate will not be affected, in addition, it is a cipher that is format compliant. However, it is not enough to achieve good security.

The proposed algorithm also permutes some coefficients of the DCT block and/or the motion vectors (MV), and this operation decreases the compression rate and then increases the bit rate. Finally, in case of performing a permutation between the non-zero DCT coefficients and those which are zero the output bit stream will not conform to the format. In [12], Bharat et al. proposed a CSC for the MPEG-1 bit stream by using the bits of the secret key to modify the sign bits of the DCT coefficients and the sign of the bits of the MV, this method should be fast, but the authors said the coding complexity is increased by about 2.55%. In addition, the process of encrypting the sign bit of DCT and MV does not achieve a high level of security.

Lian and his collaborators [13] have developed chaotic stream encryption to protect video content. This work is directed to the MPEG-2 video codec. The diagram is similar to the previous one by encrypting the signs of the DCT coefficients and the MVs. In fact, this type of encryption decreases the compression rate and can crash the decoder.

DCT Coefficients Scrambling This method encrypts the video by scrambling the c-cosine transform (DCT) into a 16 * 16 macroblock after compression. The main encryption algorithms are: Full scrambling algorithm, subsection scrambling algorithm, high-low frequency scrambling algorithm, Sub-block scrambling algorithm. The full scrambling algorithm 64 scrambles the DCT coefficients into a [14] macroblock. The non-zero coefficients are about 1/3 of the total coefficients after discrete cosine transformation. The encryption space of a macroblock is (16 * 16 * 1/3)! = 85!. However it reduces the compression ratio. The subsection scrambling algorithm splits the DCT coefficients and scrambles them into different segments based on security and compression ratio. The cipher space is $N1! * N2! \cdots$, where $N1 + N2 + \cdots = (16 * 16 * 1/3)$. The high and low frequency scrambling algorithm scrambles the DCT coefficients between high frequency and low frequency. The cipher space is $(16*16*1/3)! = 85$. This makes the coefficients very disordered and increases security. The subblock scrambling algorithm scrambles all [15] macroblocks. The cipher space is $N!$, where N is the number of macroblocks.

The DCT coefficient scrambling algorithm only changes the order of the coefficients or macroblocks and it does not encrypt the video data, it is insecure. The security of the high-low frequency scrambling algorithm is good, the subsection srambling algorithm is lower than full scrambling, and the sub-block srambling algorithm is the lowest. The srambling time is shorter than the encryption time, so the DCT coefficients scrambling algorithm is very fast, in addition, it does not touch the header information which has operability data.

2.2.2 AVC and SVC Video Encryption Algorithm

Transparent technique Magli et al. [16] proposed a perceptive (transparent) encryption technique for the AVC(H.264) and SVC, the proposed algorithms have been classified into three groups:

1. First group: encoding with drift control to decrease video quality. This can be achieved by encrypting the DCT coefficients of macro blocks (MB) I and P frames, the least significant bit (LSB) of the chosen DCT coefficients is removed by shifting to the right, then all removed bits are collected and compressed using [16] arithmetic encoding. At the end, the compressed file is encrypted using a stream cipher before sending to the decoder. One of the main disadvantages of this system is the removal of the LSB bit, as this affects the compression and the secondary information (the collected, encrypted and compressed LSB data) must be sent to the decoder, which is not preferable in the video compression and transmission.
2. Second group: used to degrade video quality, and two quality levels of I are created, one is the correctly encoded and unencrypted I frame, while the other is the bad frame (encrypted and encoded). The obvious disadvantage of this method is redundancy, and it also needs a specific decoder because it corrupts the standard decoder.
3. Third group: Enhancement Layer Encryption (SVC) using the AES algorithm. The base layer is not encrypted, so it can be used as a reference, and to decode it the user must have the secret key. As AES is used to encrypt the enhancement layer and as the author wrote, the encrypted enhancement layer does not conform to the format.

Video Scrambling by Intra Prediction Mode In [17], Ahn et al. proposed a simple and fast scrambling method for Intra prediction of AVC video. In case of Intra 16×16 MB size, only four modes are available. The Intra prediction mode is encoded using VLC, and encoded together with a model of the block encoded in luma and chroma values. The encryption process must be performed ensuring that the patterns of the encoded block do not change. A bit of the random sequence is used to encrypt the prediction mode based on: if this bit is 1, the mode is changed to the value that preserves the CBP (Coded Block Pattern), otherwise the mode is not changed. The process of modifying only I-frames, as discussed and proven, does not guarantee a high level of security. Moreover, the process of changing the prediction mode decreases the quality of the prediction process, but it does not cause a completely incorrect prediction. Finally, this method can be a technique of format-compliant security.

Encryption during entropy coding in AVC/H.264 It uses CAVLC and CABAC to accomplish the entropy coding step. Lees et al. [18], introduced a selective cipher system to encrypt the CABAC bitstream using a chaotic stream cipher based on the piecewise chaotic linear map (PWLCM). In the proposed algorithm, each binarization process has a specific encryption and decryption operation. It conforms to the format, but it affects compression and bit rate since it encrypts Unit Code (UC), Truncated Unit Code (TU), and Fixed Length Code (FLC). Finally, not all parameters during the binarization process can be encrypted while preserving the conformance property of the format.

Selective encryption based on AVC/H.264 Lian et al. [19] proposed an encryption scheme for H.264. This diagram has four parts:

1. Partial encryption of intra-forecast mode: it encrypts the code suffix term EGK (Exp-Golomb of order k) of the intra-forecast mode. The suffix term is encrypted using a [20] stream cipher.
2. Partial encryption of coefficients: DCs are encoded with VLC and then encrypted using a stream cipher proposed in [20]. While the sign of ACs is first encrypted and then encoded by VLC.
3. Partial encryption of motion vectors: encryption of the sign of the motion vector is fast and format compliant without affecting the compression ratio.
4. Key generator: All previous encryption methods in this scheme are controlled by different sub-keys. The necessary ones are generated using the key generator. Proposed system is highly efficient in terms of encryption overhead and introduces a new idea to encrypt the suffix of some values to preserve the compliant format and constant compression rate.

Fast securing AVC by selective encryption To maintain format conformance and for minimal impact on compression performance, only suffix bits encoded during CABAC entropy encoding [21]. Shahid et al. [22], proposed fast chain of access protection using selective encryption for CABAC intended for I and P images. The encryption process is performed as follows:

$$y = (x + \gamma) \quad \mod \quad \log_2(x + 1) \qquad (1)$$

where γ is given by:

$$\gamma = rand() \quad \mod \quad \log_2(x + 1) \qquad (2)$$

The parameter x is the exponential-Golomb code suffix, and x is the encrypted value of x. Later, Shahid et al. [23] enhanced this encryption scheme to encrypt the CAVLC bitstream and CABAC bitstrings of I and P frames using the Cipher Feed-Back (CFB) mode of the AES algorithm. Asghar et al. [24], present some encryptable parameters in the SVC that ensure constant bitrate and format-compliant video encryption such as UEG3 suffix (where UEG3 is a concatenation of the Unary code and the EGK code when $K = 3$), UEG0 suffix and the sign of all nonzero CTQs (Quantized Transformed Coefficients).

Fast AVC Securing by Reduced CAVLC Encryption Fast protection of AVC video content, based on selective encryption for a subset of CAVLC entropy coding is introduced by Dubois et al. [25].

This encryption scheme, like others, uses CFB-based AES to encrypt a subset of codewords/character strings of equal length during CAVLC entropy coding. Five syntaxes are used at the CAVLC entropy stage to encode the levels and the paths: coeff token, the sign of ones, of the remaining non-zero levels, of the total number of zeros and of the sequences of zeros. The cipherable parameters of this system are the sign of one and the other non-zero remaining levels to preserve the conformity of the format. Nothing new in this system except that it analyzes the level of the prediction error to decide which MB should be encrypted, but the selected encryption settings are not enough to ensure a high level of security and they affect the rate compression.

Design of new unitary transformations A new and different approach to applying
selective encryption for AVC/H.264 has been proposed by Yeung et al. [26, 27]. The
proposed algorithm selects one of multiple transformation matrices based on the bits
of the secret key. The proposed video encryption introduced a new transformation
matrix to be used instead of the DCT matrix during the transformation phase, in
addition, the sign of the DCs is included in this encryption process. It is clear that this
process requires including these alternating transformation matrices in the standard
encoder and decoder.

2.2.3   Encryption Algorithms for HEVC/H.265 Video

An encryption system has been introduced in [28] for HEVC to degrade the quality
of the video, while the full quality version is only allowed for authorized customers.
The proposed system uses encryption techniques similar to those that were proposed
in the old video standard such as: residual information sign encryption, MVD, MV
prediction index and MV benchmark. Wallendael et al. extended their previous [28]
research to identify HEVC syntax and elements that can be modified/encrypted while
preserving the compliant format for the HEVC decoder as [29]:

1. Set of short-term reference images (RPS).
2. Inter information (reference image indices, MV prediction indices, motion fusion
   indices, MVD).
3. Residual information.
4. The deblocking filter parameters.

Later, Shahid et al.[30] proposed a selective encryption scheme to secure HEVC.
Most of Shahid's work follows the same idea, which introduces a common encryption and compression system based on CABAC bin strings. Shahid looks for CTQ
sign bit, TR_p suffix, EG0, EG_1, and MVD sign bit are the inputs to the selective
encryption scheme. The sign bit encryption is straightforward, while the TR_p, EG0,
and EG_1 suffix encryption process requires special care in order to preserve format
conformance and keep the bit rate so that it is close to that obtained by compression
without encryption. Shahid defines three basic requirements for selective encryption
algorithms [30].

1. Preserve the same bitrate: the encrypted bitstream must have the same length as
   the unencrypted bitstream.
2. Maintain conforming format: the bitstream must be valid and decodable by any
   standard decoder.
3. Encrypt the dyadic cipher space: the author defines the dyadic cipher space such
   as one whose width is a multiple of 2 (i.e. 7 is not a dyadic space while 8 is
   dyadic).

The work assumes that the suffix of the (TR_p) and (EGk) codes can be encrypted
without violating any requirements, except in two [30] cases:

1. In the case where $p = 0$ in the code TR_p, since the binary representation is identical to that of the truncated unit code, and therefore TR_0 does not fulfill the first condition, consequently TR_0 is not encrypted.
2. In the case of the last equal length group of code TR_p, in this case, two binary codes are identical, whether the code EG0 is added or not, and therefore the encryption of this binary stream does not satisfy the requirement format compliance.

Encryption and decryption systems use the AES algorithm in CFB mode [31]. The encryption process is performed by bundling the bits of the suffixes EGk and TR_p to prepare the 128 cipherable plaintext bits for the AES-CFB algorithm. Below are some comments on the work of Shahid and others:

1. Residual binarization process: In their work, Shahid assume that in his algorithm residual binarization is performed using TR_p and EG0. However, in the case of HEVC as is clear in the algorithm, EG0 will never be used. Therefore, this assumption is not valid.
2. In their encryption mechanism, it is mandatory to wait until the plaintext is completely filled or the slice limit is reached. This mechanism includes an additional delay in system latency and uses more memory to fill the plaintext vector (which clearly adds more time complexity), plus the decoder works on a unit-by-unit prediction basis, and in most cases, one prediction unit is not sufficient to get a 128-bit cipherable plaintext vector. Therefore, more memory is needed and more time. Staying in the same context as the encryption work done for HEVC/H.265 video, Sallam et al. [32] have introduced a new high-efficiency video encryption technique (HEVC) and Partial Encryption (SE) for encryption of highly sensitive video bitstream data. The SE-HEVC technique must maintain video format conformance, must have the same bit rate, and ensure real-time constraints. These characteristics result from the use of low computational complexity RC6 block ciphers for selective encryption. The proposed SE-HEVC-RC6 encrypts the sign bit of the discrete cosine transform (DCT) coefficients, the suffixes of the remaining absolute values DCT which are binarized by Exp-Golomb (EGk) of order zero, the sign bits of the motion vector difference (MVD) and the suffixes of the absolute values MVD which are binarized by EGk of order 1.

2.2.4 Chaos-Based Encryption Algorithm

A new method for selectively encrypting sensitive data of the latest video coding standard, called High Efficiency Video Coding (HEVC) has been proposed by Ahmed et al. [33]. The proposed HEVC video selective encryption technique uses the low-complexity chaotic logistic map (CLM) to encrypt the motion vector difference (MVD) sign bits and the discrete cosine transform (DCT) coefficients at the stage of the entropy coding of the video. The contribution of the CLM-based SE-HEVC proposal is to encrypt sensitive video bits with the characteristics of low complexity, fast encoding time, maintaining constant bit rate and HEVC format compliance.

A selective encryption with minimal algorithmic complexity intended for public networks has been proposed by Lui and Wong [34], it is a chaos-based selective encryption scheme implemented on the H.264/AVC standard. The scheme uses multiple chaotic Rényi maps to generate a pseudo-random bit sequence that is used to mask selected H.264/AVC syntax elements. It provides sufficient protection against a full rebuild while maintaining the conformance property of the format so as not to cause a decoding error without the key. Operational efficiency is high due to the low computational complexity of Rényi's chaotic map. Moreover, this algorithm is very sensitive to the secret key and has good perceptual security.

In the research work proposed by Altaf et al. [35], the requirements of reduced computing with increased security, format conformance and compression efficiency are addressed. The security problem is solved by carefully selecting the override boxes for the block cipher used. For compression efficiency and format conformance, video data is selected so that statistical and structural characteristics are preserved. In order to increase security, various chaos-based substitution boxes, which are an integral part and the only nonlinear operation of block ciphers, have been investigated for cryptographic attacks. The selected substitution boxes were used for the swapping of the selected video data and its encryption by integrating the S-box with the advanced compression standard and H.264/AVC. The video data selected to be secure consists of discrete cosine transform coefficients; the signs of these non-zero coefficients and transformation coefficients. The discrete cosine transform coefficients were permuted using the selected S-box, while the signs of these nonzero coefficients were fully permuted using AES with a modified S-box table. This algorithm conforms to the format and has a reduced computation time.

Ganeshkumar et al. [36], proposed a chaos-based three-level secure cryptography system for video encryption. The proposed method uses a round permutation and a broadcast structure for encryption. Logistic map and Tent map are combined (LTS) to generate the initial parameters of the proposed encryption system. The video frame is first selected based on the frame selection (FS), and then it is encrypted by applying the swap order (PO). Finally, the broadcast is performed on the permuted frame by generated broadcast (DF). After encryption, the encrypted frame is placed in order to create a video sequence for transmission. This method offers three-level security (FS, PO and DF) for the extraction of encrypted information. The security analysis and the results obtained by this encryption confirm the robustness and competence of this method.

Chandrasekaran et al. [37], also propose a research work that aims to improve the speed of naive approaches by designing chaos-based S-boxes. Chaotic equations are generally known for their randomness, extreme sensitivity to initial conditions, and ergodicity. The proposed methodology uses a two-dimensional discrete Henon map for the generation of a dynamic, key-dependent S-box that could be integrated with symmetric algorithms like Blowfish and Data Encryption Standard (DES) and disposable mask algorithms based on the generation of unique keys for simple substitution encryption. This approach confirms that chaos-based S-box design and key generation significantly reduce video encryption computational cost without any compromise on security.

A selective encryption system for High Efficiency Video Coding (HEVC) has been proposed by Peng et al. [38], to enhance privacy information in the videos, In this approach, a pseudo-random binary sequence is first constructed by a key generator based on the chaotic Rossler system. Then, the generated keystream is used to encrypt the movement information and residual coefficients of HEVC. In particular for the ciphering of the improved scrambling effect is obtained by ciphered DC coefficients. It has been proven by the security analysis that this cryptosystem can produce a good encryption effect, and keep the encrypted stream conforming to the format.

2.2.5 Synthesis

This section in the form of a state of the art presents certain approaches used to secure a video stream, and gives a comparison between these different techniques, specifying certain advantages and disadvantages for each of them. Full security algorithms are the most secure in terms of security, but they are not suitable for applications that operate in real time since they have a higher computation time compared to encryption selective. Thus, selective security algorithms on the other side can satisfy this constraint in real time, although they present a slightly lower level of security compared to full security since they do not secure all video data. To conclude, encrypting a streaming video stream is a trade-off between security and real-time performance.

3 Proposed Chaos-Based Video Cryptosystem

The proposed cryptosystem consists of two essential parts, the chaotic key generator, which generates the encryption keys, and the encryption algorithm, which encrypts the video frames according to the steps shown in Fig. 1.

3.1 Chaotic Keys Generator

The chaotic key generator is based on Chen's chaotic system defined by the following equations:

$$\begin{cases} \frac{dx}{dt} = \sigma(y - x) \\ \frac{dy}{dt} = (\rho - a - y)x + \rho z \\ \frac{dz}{dt} = xy - \beta z \end{cases} \tag{3}$$

where $\rho = 28$, $\sigma = 35$, $\beta = 3$.

The previous system is a nonlinear system, so to solve it we use a numerical resolution method, which is the Rung-Kutta method, then by applying a transformation

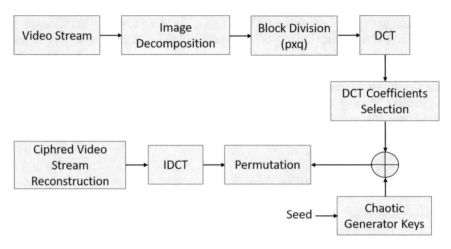

Fig. 1 Encryption process

algorithm to the solutions obtained, we manage to transform the part fractionalization of these solutions into a statistically random binary sequence consisting of zeros and ones. In the following, we will detail the key generation algorithm.

3.1.1 Proposed Ciphering Keys Generation Algorithm

The following algorithm assumes that the image to be encrypted is of RGB type and of dimensions n and m. However, it is not the whole image that will have to be encrypted but only a few pixels, which will be selected by a selection algorithm. We then note N_c the number of these selected pixels, the explanation of the algorithm details is illustrated by the flowchart of Fig. 2 and explained by the following steps:

1. Firstly, solve the chaotic system of Chen by Runge-Kutta for an interval $t \in [0, 0.01 \times N_c]$ with a step $dt = 0.01$;
2. Secondly, concatenate the three obtained solutions x, y and z into a single vector $S = \{S^{(1)}, S^{(2)}, \ldots, S^{(3 \times N_c)}\}$;
3. Calculate the vector $F = \{F^{(1)}, F^{(2)}, \ldots F^{(3 \times N_c)}\}$, which is the fractional part of S where:

$$F = fract(S) \tag{4}$$

However, the fractional part of an element $F^{(j)}$ belonging to the vector F taking 8 decimal places is written as follows:
$$\forall j \in [0, 3 \times N_c];$$

$$F^{(j)} = 0.f_1^{(j)} f_2^{(j)} f_3^{(j)} f_4^{(j)} f_5^{(j)} f_6^{(j)} f_7^{(j)} f_8^{(j)} f_9^{(j)} f_{10}^{(j)} f_{11}^{(j)} \tag{5}$$

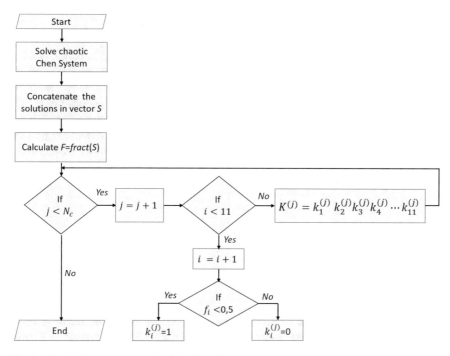

Fig. 2 Chaotic keys generator algorithm flowchart

4. Finally, a post-processing algorithm is applied on the fractional part to obtain the encryption key $K = \{K^{(1)}, K^{(2)}, \ldots K^{(3 \times N_c)}\}$ so that it acquires better statistical properties.

Algorithm 1 Post-processing algorithm

For $j = 0$ to N_c do
For $i = 0$ to 11 do

$$k_i^{(j)} = \begin{cases} 1 \ si \ f_i^{(j)} < 0.5 \\ 0 \ si \ f_i^{(j)} >= 0.5 \end{cases} \tag{6}$$

End
$K^{(j)} = k_1^{(j)} k_2^{(j)} k_3^{(j)} k_4^{(j)} k_5^{(j)} k_6^{(j)} k_7^{(j)} k_8^{(j)} k_9^{(j)} k_{10}^{(j)} k_{11}^{(j)}$
End

Generator Perturbation

The generator must never return to the same initial state so as not to regenerate the same keys. Another secret parameter 's' is therefore added to the initial conditions of this generator to modify its behavior, because it is very sensitive to the initial conditions, but only a slight disturbance must be applied so that it does not go out of its regime chaotic. Indeed, a test on MATLAB was carried out by generating a key k_1 with the initial conditions x_0, y_0, and z_0, then by changing a single digit after the decimal point of these initial conditions, we obtain another key k_2 completely different from the first key k_1, with a relative deference of 99.96% between the two keys. The initial conditions of this generator will be modified as follows:

$$\begin{cases} x_0 = x_0 + s \times 10^{-11} \\ y_0 = y_0 + s \times 10^{-11} \ with: 0 \le s \le 10^{10} \\ z_0 = z_0 + s \times 10^{-11} \end{cases}$$

3.1.2   Key Generator Evaluation

The advantage of this generator compared to others is that it allows to have as many zeros as ones, because it always compares with 0.5, moreover, it facilitates the passage to the binary representation directly for the languages of the high level without going through the decimal representation. The evaluation of the generator consists in carrying out statistical tests on the sequences produced by this generator in order to determine whether they are random or not. Thus, the statistical tests of the NIST battery were applied to 100 sequences of 1 million bits. By downloading the test battery *sts2.1.2* from the official site [39], we have obtained the results summarized in Table 1.

The sequences produced by this generator pass the NIST test if the proportion for each test is greater than 96%, except for Random Exursion Variant and Random Exursion, which must be greater than 95%, also the p_T value must be greater than 10^{-4} for each test in order for the sequence to pass that test. The value of p_T gives us an idea of the distribution of the sequences, so that a value of $p_T > 10^{-4}$ indicates that these binary sequences are evenly distributed. So, according to the Table 1, all the sequences tested pass the test, and the subkey values produced by this generator are uniformly distributed as shown in Fig. 3, this uniformity is tested Using MATLAB by plotting the histogram of the encryption key for each frame, this key is the result of the concatenation of the subkeys that represent the solution of Chen's chaotic system in instant t. The uniformity of these 11-bit subkeys also proves that the binary sequences generated by this chaotic generator are statistically random and that they can therefore be used to construct the encryption keys. Indeed, it is almost impossible for an attacker to find the encryption keys by statistical analysis or by differential analysis because they have particular statistical properties which add robustness and increase the security of the proposed encryption system.

Table 1 NIST statistical test results

Statistical test	P-value	Proportion	Result
Frequency	0.911413	1	Passe
Block frequency	0.000477	1	Passe
Cumulative sums	0.779188	1	Passe
Runs	0.816537	0.97	Passe
Longest run	0.554420	0.98	Passe
Rank	0.924076	0.98	Passe
FFT	0.181557	0.97	Passe
Non overlapping template	0.145326	1	Passe
Overlapping template	0.171867	0.99	Passe
Universal	0.883171	0.99	Passe
Approximate entropy	0.816537	0.99	Passe
Random excursions	0.654467	1	Passe
Random excursions variant	0.002175	1	Passe
Serial	0.657933	0.98	Passe
Linear complexity	0.171867	1	Passe

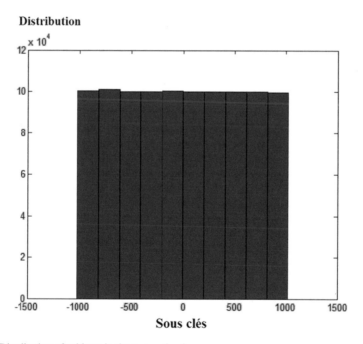

Fig. 3 Distribution of subkeys in the encryption key

3.2 Encryption/Decryption Algorithm

The proposed encryption algorithm is designed for frame-by-frame video encryption, the video is therefore considered as a sequence of images separated in time at least by $\Delta t = 0.04$ s which corresponds to 25 frames per second, the minimum frequency for which the eye perceives video smoothly. As shown in Fig. 1, this algorithm relies on data compression by performing a frequency transformation (DCT) on blocks of size $p \times q$, then the DCT coefficients are selected and xored with the key. Finally, by performing a pseudo-random permutation and returning to the spatial domain using the inverse discrete cosine transform (IDCT) the image will be encrypted. For the decryption, we proceed by the inverse operation, that is to say that we perform the permutation after the calculation of the DCT, then we xorize with the key, then the inverse transformation of the DCT is carried out to recover the original picture. However, the compression performed in this algorithm is lossy compression, so on reception the original image is recovered with some visual degradation which depends on the bit rate of the video transmission. This relationship will be explained in detail later in this chapter.

3.2.1 Transformation Fréquentielle (DCT)

DCT is a frequency domain transformation used to perform data compression by exploiting the fact that human eye perceives low frequencies that give structure and form of an image much more than the high frequencies which contain the details of the image, so the basic idea of this type of compression is to eliminate spatial redundancy by eliminating the high frequencies and not keeping than low frequencies.

Figure 4 shows the DCT transformation principle and Fig. 5 shows the position of pixels in a block after the DCT transformation. This transformation rearranges the pixels so that the first coefficient at the upper left end of the block is called DC, which has the highest value but the lowest frequency. Starting from this coefficient, descending and moving away towards the right of each block, the frequency increases and the value of the coefficients decreases in absolute value until having zero terms at the lower right end of the block.

The best and most efficient way to compute the discrete cosine transform of an image is to break it down into N blocks of size $p \times p$, as the H.264 standard does which performs a division into 8×8 blocks and then applies the DCT for each block using a transformation matrix T, as indicated in by relation (7). Indeed, the matrix makes it possible to calculate the coefficients only once, then to save them in memory and to use them directly without recalculating them each time, which reduces the calculation time. Also, there are MATLAB or Python programs with optimized matrix calculation algorithms that serve to further reduce the time of this matrix calculation.

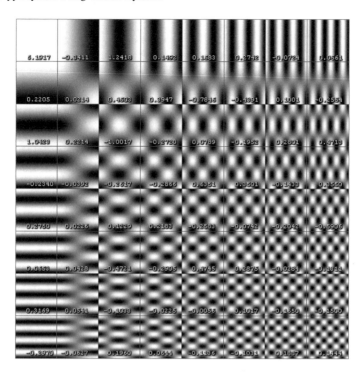

Fig. 4 DCT transformation

Fig. 5 DCT coefficients position

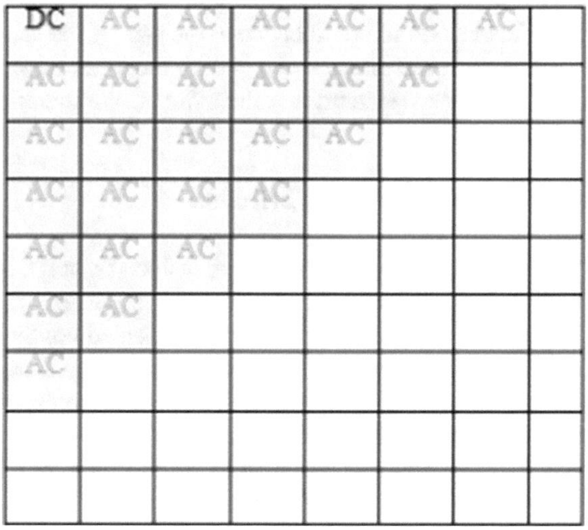

DC	AC	AC	AC	AC	AC	AC	
AC	AC	AC	AC	AC	AC		
AC	AC	AC	AC	AC			
AC	AC	AC	AC				
AC	AC	AC					
AC	AC						
AC							

$$T_{ij} = \begin{cases} \frac{1}{\sqrt{B}} & if\ i = 0 \\ \frac{2}{\sqrt{B}} \cos(\frac{(2j+1)i\pi}{2N}) & if\ i > 0 \end{cases} \tag{7}$$

$$DCT(Image) = T_p * bloc_i * T_p^t \qquad i \in \{1, 2 \dots N\} \tag{8}$$

where T_p is the transformation matrix of dimension $p \times p$, B is the size of each block and N is the total number of blocks. However, the image is not square, it often has a size of $n \times m$ with $n \neq m$, in this case it is more optimal to decompose it into rectangular blocks of dimensions p and q to traverse more space using the minimum number of blocks, then relation (8) becomes:

$$DCT(Image) = T_p * bloc_i * T_q^t \qquad i \in \{1, 2 \dots N\} \tag{9}$$

With T_p and T_q are the square DCT transformation matrices of dimensions $p \times p$ and $q \times q$ respectively.

Dividing the image into multiple blocks and processing each block separately is more convenient, since it reduces memory usage and allows better data control, but can also increase computation time because the same operation is repeated as many times as the number of blocks. For example, dividing the "Lena" image on Matlab into 8×8 blocks (1024 blocks) takes 1 s to compute the DCT transformation, while it only takes 0.3 s if you divide it into 64×64 (256 blocks). The quality of the compression is not affected by the number of blocks selected, since this quality only depends on the ratio between the size of the selected block and the number of coefficients taken into account in each block, as shown later in Eq. (14). In short, the more the block size increases, the more the computation time of the DCT is reduced.

3.2.2 DCT Coefficients Selection

The mask has the same block dimensions $(p \times q)$, and the index k represents the maximum number of 1s in a row or column. As the arrangement of the 1s in the mask M_k is triangular, the total number of selected coefficients in a block is as follows: The most important DCT coefficients are located on the left side at the top of the block, these are the ones that must be kept while the others are close to zero, so their effect is neglected in order to compress the image by rounding to zero all values less than 0.01. By traversing each block with the 'zig-zag' technique illustrated by Fig. 6, it is the same as multiplying it by a triangular mask (matrix (11)) containing 1s in the region we want preserve and zeros in the region you want to ignore. Thus, by carrying out this elimination the image will be compressed and it will be full of zeros (hollow) that we denote Image'. Then, to select the DCT coefficients which will be encrypted thereafter, it suffices to perform an element-by-element multiplication of the matrix obtained from relation (9) and the mask defined by (11).

Fig. 6 DCT coefficients selection by the zig-zag technique

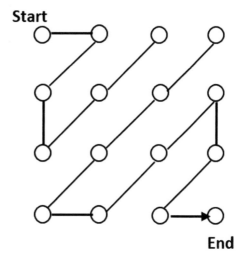

$$Image' = M \otimes (T_p * bloc_i * T_q^t) \qquad i \in \{1, 2...N\} \tag{10}$$

$$M_k = \begin{pmatrix} 1 & \cdots & 1 & 0 & \cdots & 0 \\ \vdots & \ddots & \ddots & \ddots & \ddots & \vdots \\ 1 & & \ddots & \ddots & \ddots & \vdots \\ 0 & & \ddots & \ddots & \ddots & \vdots \\ \vdots & & \ddots & \ddots & \ddots & \vdots \\ 0 & \cdots & & & \cdots & 0 \end{pmatrix} \tag{11}$$

The mask has the same block dimensions $(p \times q)$, and the index k represents the maximum number of 1s in a row or column. Since the layout of the 1s in the mask M_k is triangular, the total number of coefficients selected in a block is thus:

$$Nc = k + (k-1) + (k-2) + ...1 \qquad \forall \; 0 \leq k \leq p \tag{12}$$

Thus:

$$N_c = \frac{k(k+1)}{2} \qquad \forall \; 0 \leq k \leq p \tag{13}$$

Then the compression rate CR which is defined as the ratio between the size of the original image $(n \times m)$ and the sum of the number of pixels taken into account N_c for the N blocks of size $p \times q$ will be equal to:

$$CR = \frac{n \times m}{N_c} = \frac{p \times q \times N}{N \times N_c} = \frac{p \times q}{N_c} \tag{14}$$

Thus:

$$CR(k) = \frac{2 \times p \times q}{k(k+1)} \tag{15}$$

Formula (15) shows that the compression rate is a function of k, which allows it to vary according to the value of this parameter k. In addition, in the transmission of video, the bit rate must always be constant and limited, moreover, it is always necessary to keep a $\Delta t = cte$ between two successive images to have a constant number of images per second in order to always remain within the standard used. For example, in the PAL standard $\Delta t = 0.04$ s corresponds to 25 frames per second, so from these two conditions, the video must be compressed by calculating the value of $k_{\max}$ which allows to satisfy the above constraints. Therefore, we try to optimize k by modeling these constraints as follows:

$$\begin{cases} Debit \geq Nb_{(bits/s)} \\ \\ \Delta t = \frac{1}{Nb_{Images/s}} \end{cases} \tag{16}$$

where $Nb_{(bits/s)}$ is the number of bits selected in an image to be encrypted and transmitted in one second, and is equal to N_c times the number of blocks N, multiplied by the number of images per second $\frac{1}{\Delta t}$, multiplied by the number of bits used to encode a single pixel in an RGB image, so:

$$Nb_{bits/s} = 8 \times 3 \times \frac{1}{\Delta t} \frac{k(k+1)}{2}.N \tag{17}$$

Therefore:

$$Nb_{bits/s} = 12 \frac{k(k+1)}{\Delta t}.N \tag{18}$$

The system of Eq. (16) then becomes:

$$Nb_{bits/s} = 12 \frac{k(k+1)}{\Delta t}.N \tag{19}$$

Thus:

$$Debit = 12 \frac{k_{\max}^2 + k_{\max}}{\Delta t}.N \tag{20}$$

We therefore obtain the following quadratic equation to be solved:

$$k_{\max}^2 + k_{\max} - \frac{\Delta t \times Debit}{12N} = 0 \tag{21}$$

The solution of the previous equation allows to find the value of $k_{\max}$, for an integer $k_{\max}$ between 0 and p.

$$k_{\max} = E \left[\frac{\sqrt{\frac{Debit \times \Delta t}{3N} + 1} - 1}{2} \right] \quad (22)$$

By replacing (20) by (15), we obtain the formula allowing to vary the Compression Rate (CR) according to the value of the bit rate and the number of images per second according to the standard used.

$$CR = 24 \frac{Nb_{Images/s} \times p \times q}{Debit} . N \quad (23)$$

3.2.3   Permutation

After selecting the DCT coefficients, we only change the position of the AC coefficients in each block without touching the location of the DC coefficient, because the DC coefficient is very predictable, so there is no need to change its position. The switching algorithm must be simple but robust so as not to increase the complexity of the algorithm because it is a real-time video transmission on the one hand, and on the other hand to add more robustness to the cryptosystem in order to ensure more dissemination. In fact, since the image is full of zeros due to compression, selective encryption of DCT coefficients in the frequency domain is not sufficient to remove the correlation between adjacent pixels when returning to the spatial domain, so that changing the location of the DCT coefficients before applying the inverse transform is too necessary to remove the correlation between adjacent pixels in the encrypted image. Let K be the encryption key and v the value of the DCT coefficients, the permutation algorithm used inspired by $RC4$ is described as follows:

Algorithm 2 Permutation Algorithme

For $i = 0$ to N_c
$j = i + j + K^{(j)}$ mod N_c;
permute v_i and v_j ;
End

3.2.4   Encryption Operation

The last step of this algorithm consists of performing a *xor* between the different blocks of the image and the key generated by the chaotic generator, then all that remains is to group the different images and return to the spatial domain by applying the transform inverse IDCT (relation (24)) to have the encrypted video. The encryption is based on the One Time Pad (OTP) cipher, where the encryption keys should only be used once to encrypt the data. In this algorithm, each frame of the video

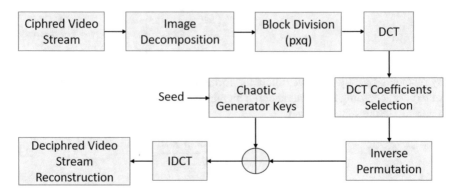

Fig. 7 Decryption process

sequence is therefore encrypted by a completely different key in order to guarantee that an attacker can never find the key by the attack with a plaintext.

3.2.5 Decryption Operation

After the encryption of the video stream, it would be quantized and encoded and then transmitted to the receiver where it would be decrypted by performing the reverse operation of the decryption algorithm as shown in Fig. 7.

After recovering the encrypted stream, the DCT is calculated, reverse permutations are performed to recover the original location of the DCT coefficients and finally, decryption is done by an xor with the encryption keys. Just after, the inverse discrete cosine transform is applied to the reconstructed image to return to the spatial domain and recover the positions and values of the original pixels, all that remains is to regroup these images to reconstruct the video in clear.

$$IDCT(Image) = T_p^t * bloc_i * T_q, \qquad i \in \{1, 2...N\} \tag{24}$$

4 Performances Evaluation

For the evaluation of the performance of the cryptosystem explained above, it is based on three main criteria, namely the degradation of the video quality after encryption, the analysis of the security of the encrypted video using measurements based on the mathematical study and time evaluation by studying the time required for encryption after the optimization of the algorithm to achieve real-time operability.

Table 2 Reference videos

Video sequence	Resolution	Image/s
Jockey	1920 × 1080	120
Four people	1280 × 720	60
Forman	352 × 288	30

Table 3 Mean PSNR values

Method	Video sequence	PSNR (dB)
Proposed cryptosystem	Jockey	8.62
Sallam et al. [32]	Jokey	8.63
Sallam et al. [33]	Jokey	10.67

Table 4 Mean SSIM values

Method	Video sequence	SSIM
Proposed cryptosystem	Jockey	0.007
Sallam et al. [32]	Jokey	0.02
Sallam et al. [33]	Jokey	0.029

4.1 Assessing Video Quality Degradation

Safety quality assessment involves the use of quality measurement metrics such as PSNR and SSIM, while these metrics are much more used to assess video quality after decompression, but they can also be used to measure video degradation after encryption.

For properly secured video, the value of PSNR and SSIM should be minimal. Tables 3 and 4 show the average values of PSNR and SSIM by comparing the proposed cryptosystem to other reference ones, and by using the same video sequences (Table 2) in order to establish a fair comparison. We notice that the PSNR and SSIM values obtained are low, which means that the degradation of the Video Jockey at the encryption level is quite significant, so that the cryptosystem makes the information contained in the video stream incomprehensible and inaccessible to a person unauthorized.

4.2 Security Analysis

In this section, in order to evaluate the performance of the proposed cryprosystem in terms of security robustness, we perform an in-depth security analysis using the most adequate performance criteria.

Fig. 8 Clear image #5 from
Video Jockey

Table 5 Key sensitivity results

Test	Value
NPCR (%)	99.6104
UACI (%)	33.1873
HD (%)	0.4993

4.2.1   Keys Space

The encryption key changes for each frame, while the secret key remains fixed for the entire transmission. In this case, the attacker is not focusing on the encryption keys themselves, but on the chaotic key generator trying to find his six secret parameters from Chen's chaotic system: ρ, σ, β and the initial conditions x_0, y_0, z_0. The fractional part of each of these parameters is 44 bits, so the key space is equal to $(2^{44})^6 = 2^{264}$, which is greater than 2^{128}, so we can say that the brute force attack is useless for this key generator.

4.2.2   Key Sensitivity

Key sensitivity is measured by applying the metrics NPCR, UACI and HD to two frames encrypted by a slightly modified one-bit key, Figs. 9 and 10 are the encrypted frames of the same frame #5 of the joker video (Fig. 8). We slightly disturb the secret parameter σ of the chaotic key generator, the results of the key sensitivity test are presented in Table 5. This result was expected since one of the main properties of chaotic systems is sensitivity to initial conditions. Thus, by slightly modifying the σ secret parameter, this will generate a completely different encryption key and therefore two completely separate encrypted images.

Fig. 9 Encrypted Image #5
of Video Jockey with
$\sigma = 36.578476337452983.$

Fig. 10 Encrypted image #5
of Video Jockey with
$\sigma = 36.578476327452982$

4.2.3   Differential Attack Analysis

Otherwise known as a plaintext attack, this attack consists of comparing two plaintext images at the input of a cryptosystem and then comparing their encrypted ones at the output. In order to measure the robustness of the proposed cryptosystem against this type of attack, certain metrics are used such as the NPCR, UACI and the Hamming distance. Tables 6, 7 and 8 show the obtained results compared to other references.

The optimal value of the NPCR is 99.6094%, that of the UACI is 33.4635% and the Hamming distance is 50%. The comparison of the obtained results with the optimal values and the reference results therefore shows that the proposed cryptosystem is robust against plaintext attacks.

Table 6 NPCR results

Method	Image	NPCR (%)
Proposed cryptosystem	Image #50 of four people video	99.62
Sallam et al. [32]	Image #50 of four people video	99.71
Proposed cryptosystem	Lena 256 × 256	99.61
Farjallah [22]	Lena 256 × 256	99.58

Table 7 UACI results

Method	Image	UACI (%)
Proposed cryptosystem	vidéo de four people image #50	33.24
Sallam et al. [32]	vidéo de four people image #50	38.23
Proposed cryptosystem	Lena 256 × 256	33.19
Farjallah [22]	Lena 256 × 256	33.44

Table 8 Hamming distance results

Method	Video sequence	HD (%)
Proposed cryptosystem	Lena 256 × 256	50.218
Farjallah [22]	Lena 256 × 256	49.990

4.2.4 Statistical Analysis

Statistical analysis ensures that the image frames encrypted by the proposed cryptosystem have statistical properties that render cryptanalysis that attempts to reveal patterns and plaintext image information unnecessary. There are several tools of this analysis, in the following we present the most important ones.

Histogram The histogram gives the visual distribution of the pixels on the color levels (255 levels for an image coded on 8 bits), while the Chi-square function makes it possible to quantify more precisely this distribution of the pixels. Figures 13 and 14 are the histograms of the clear (Fig. 11) and encrypted (Fig. 12) image #50 of the Four people video respectively, and Table 9 indicates the Chi-square value calculated for the proposed cryptosystem and the value calculated for a reference cryptosystem. For the image to be properly secured, its histogram must be uniform, as shown in the Fig. 14, so the calculated Chi-square value must be less than 293, which is the case for the proposed cryptosystem, as shown in Table 9.

Correlation The correlation between adjacent pixels according to the horizontal, vertical and diagonal direction must be minimal for an encrypted image, as it is shown by Table 10 where the correlation according to the three directions approaches zero for the proposed cryptosystem and others references. Figures 15, 17 and 19 represent

Table 9 Chi-square results

Method	Video sequence	Chi-square
Proposed cryptosystem	Lena 256 × 256	272.01
Farjallah [22]	Lena 256 × 256	241.90

Fig. 11 Clear image #50 of four people video

Fig. 12 Ciphred Image #50 of four people video

the correlation between adjacent pixels in the three directions for the clear image #30 of the Format video, where there is a strong correlation between these pixels seen that they are aligned on a line, although, we notice a weak correlation in the encrypted image (Figs. 16, 18 and 20) given that the distribution of pixels is in all directions.

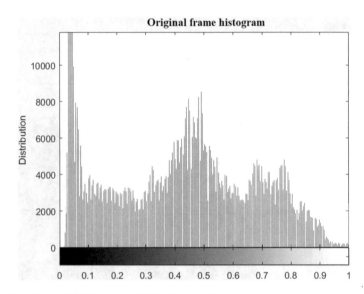

Fig. 13 Clear image #50 histogram of video four people

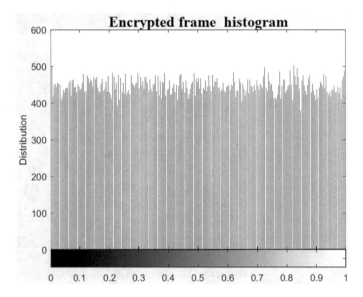

Fig. 14 Ciphered image #50 histogram of video four people

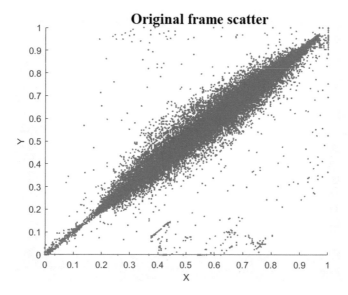

Fig. 15 Clear image #30 horizontal correlation of Forman video

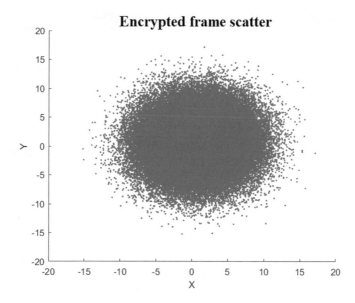

Fig. 16 Ciphered image #30 horizontal correlation of Forman video

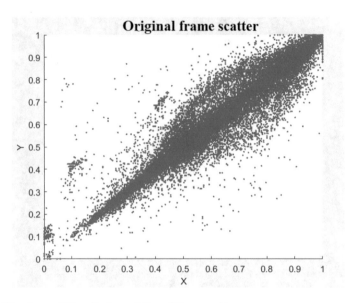

Fig. 17 Clear image #30 vertical correlation of Forman video

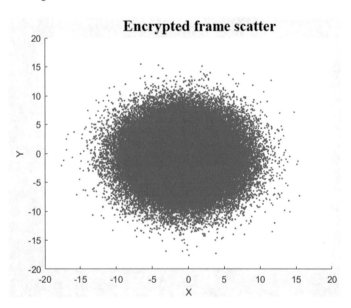

Fig. 18 Ciphered image #30 vertical correlation of Forman video

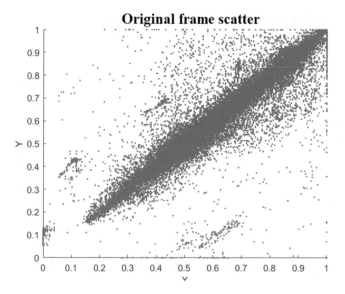

Fig. 19 Clear image #30 diagonal correlation of Forman video

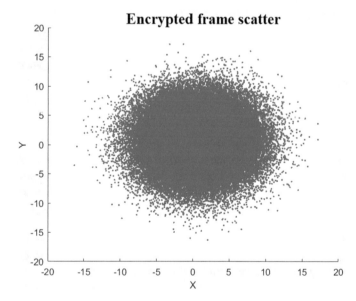

Fig. 20 Ciphered image #30 diagonal correlation of Forman video

Table 10 Results of horizontal, vertical and diagonal correlation

Method	Image	Corr H	Corr V	Corr D
Proposed cryptosystem	Image #50 of four people video	0.0321	−0.0061	0.0176
Sallam et al. [32]	Image #50 of four people video	0.0312	0.0750	0.0428
Proposed cryptosystem	Image #10 of Jockey video	0.0070	−0.0044	0.0037
Sallam et al. [33]	Image #10 of Jockey video	0.0365	0.0304	0.0994

Table 11 Entropy results

Method	Image	Entropy (bits)
Proposed cryptosystem	Image #50 of four people video	7.9915
Sallam et al. [32]	Image #50 of four people video	7.5718
Proposed cryptosystem	Image #10 of Jockey video	7.997
Sallam et al. [33]	Image #10 of Jockey video	7.460

Entropy Entropy measures the degree of disorder in an image, the entropy value for a properly encrypted image should be close to 8 bits, so comparing the entropy value for the proposed cryptosystem to the reference one (Table 11), we decide that the cryptosystem is sufficiently secure.

4.2.5 Encryption Quality

The encryption quality compares the number of occurrences of color levels between clear and encrypted images, the higher the EQ value, the more secure the cryptosystem. However, this value cannot increase infinitely. Indeed, it is limited by a maximum value representing the ideal value. Thus, the image is correctly encrypted if the EQ value is close to EQ_{max}. Table 12 summarizes encryption quality values obtained by the proposed cryptosystem and compared to those of some benchmarks. According to results in this table, the EQ value is large enough and close to the maximum value reflecting a good encryption quality.

4.3 Temporal Performances

Temporal performance evaluation is necessary for systems that operate in real time. The most time-consuming operation is key generation, so to minimize encryption time, it is necessary to generate keys of small sizes. To this end, dividing the images

Table 12 Encryption quality results

Method	Image	EQ	EQ$_{max}$
Proposed cryptosystem	Image #50 of four people video	19558	21515
Sallam et al. [32]	Image #50 of four people video	12819	21515
Proposed cryptosystem	Jockey image #10 video	14690	16137
Sallam et al. [33]	Jockey image #10 video	12762	16137

Table 13 Temporal performances

Method	Video sequence	ET (Mbps)	Ciphering time (s)	Cycle/Octet
Proposed cryptosystem	Four people video	110.59	0.2	22.60
Sallam et al.[32]	Four people video	0.011	248	$3.4.10^5$
Sallam et al. [33]	vidéo de Four people	0.18	125	$1.45.10^5$

into blocks and then generating the keys for each block, taking into consideration only a few DCT coefficients for them to be encrypted, can significantly reduce the key generation time. At the beginning an array is generated which contains a number of keys, then during encryption, with each new frame only part of the key is generated so that when the last key in the array is reached the new key would be complete, then an already used key is overwritten and replaced by the new key, this way the key matrix will be changed dynamically, although some keys will be reused, but this is a trade-off between security and real-time performance.

The time performance tests were carried out on a PC of i5-8400H 2.5 Ghz processor and 16 GB RAM. Table 13 shows that the proposed cryptosystem can achieve good real-time performance since it has an encryption throughput (ET), a lower encryption time and a lower number of cycles per byte compared to benchmark cryptosystems.

5   Synthesis

Based on the security analysis and the obtained experimental results by applying security measures, and comparing to some referenced works, the competence and the robustness of the proposed cryptosystem is proven.

Concerning temporal performances, we can note according to the Table 13, that values obtained by the proposed cryptosystem are very far from those obtained by reference works. This result was predictable considering in reference works, they perform all compression steps while the proposed approach limits the cryptosystem to the frequency transformation step to gain more in terms of computation time. However, an incomplete compression increases the bit rate. Thereby, the formula (23) is used to control the latter, since the main goal of the proposed algorithm is to achieve the real-time performance and to compress video. However, by keeping a bit-rate constant video quality is degraded. Therefore, it is a trade-off between quality and real-time operability. In addition, the encryption algorithms used in these reference systems are more complicated than the proposed encryption system. A. Sallam et al. [32] use for example the R_6 which as a block cipher which can create more latency during the encryption operation. The presented encryption algorithm is based on simple operations: disposable mask, fast permutation and optimization of key generation allowing to obtain better temporal performance in terms of data processing and encryption time.

The encryption approaches proposed in [32, 33] conform to the format since they do not encrypt DCT coefficients. On the other hand, the proposed approach does not conform to the format, because the priority is to obtain a high level of security and not to respect the used compression standard. Thus, we will not be able to decode the video properly if we perform full H.264 compression. Therefore, video encoding and decoding require an another architecture compatible only with the used encryption approach.

6 Conclusion

This chapter evaluates performance of a video stream encryption for the evaluative study of cryptosystems intended to secure the video stream transmission, and falling within the framework of the security of communications and the protection of sensitive information that must be apprehended only by the two communicating entities. The problem we encountered while designing video cryptosystem is that we cannot achieve the real-time performance we want, because the video will be very slow if fully encrypted. Thus, selective encryption has been implemented to solve this problem through the proposition of chaotic video cryptosystem. In this way, selective encryption consists of encrypting only part of the video to save computation time. At this point, the discrete cosine transform has been used to reduce the amount of encrypted data. Therefore, frames in the video stream will be split into blocks of varying lengths and processed separately. The DCT coefficients are calculated for each of them and encrypted by xoring them with chaotic keys generated using the Chen's chaotic system. A $RC4$ inspired permutation of the DCT coefficients is also added for more security. In addition to the simplicity of the previous encryption algorithm (DCT, disposable mask and simple permutation), the code responsible for chaotic key generation is optimized to reduce computation time to improve timing performance. After applying the performance criteria on the used chaotic encryption, obtained results proved the robustness and the efficiency of the presented cryptosystem in terms of security and real-time operability.

References

1. Sze V, Budagavi M, Sullivan GJ (2014) High efficiency video coding (HEVC). In: Integrated circuit and systems, algorithms and architectures. Springer, pp 1–2
2. Richardson IE (2004) H.264 and MPEG-4 video compression: video coding for next-generation multimedia. Wiley
3. Stinson DR (1995) Classical cryptography. Cryptography, theory and practice, 2nd edn. In: Rosen KH (ed) Chapman & Hall/CRC, pp 1–20
4. Qiao L, Nahrstedt K et al (1997) A new algorithm for MPEG video encryption. In: Proceedings of first international conference on imaging science system and technology, pp 21–29
5. Spanos GA, Maples TB (1995) Performance study of a selective encryption scheme for the security of networked, real-time video. In: Proceedings of fourth international conference on computer communications and networks-IC3N'95. IEEE, pp 2–10
6. Agi I, Gong L (1996) An empirical study of secure MPEG video transmissions. In: Proceedings of internet society symposium on network and distributed systems security. IEEE, pp 137–144
7. Meyer J, Gadegast F (1995) Security mechanisms for multimedia data with the example of mpeg-1 video. Université technique de Berlin, Allemagne, Description du projet SECMPEG
8. Tang L (1997) Methods for encrypting and decrypting MPEG video data efficiently. In: Proceedings of the fourth ACM international conference on multimedia, pp 219–229
9. Qiao L, Nahrstedt K (1998) Comparison of MPEG encryption algorithms. Comput Graph 4:437–448
10. Tosun AS, Feng W-C (2000) Efficient multi-layer coding and encryption of MPEG video streams. In: 2000 IEEE international conference on multimedia and expo. ICME2000. Proceedings. Latest advances in the fast changing world of multimedia (cat. no. 00TH8532). IEEE, pp 119–122
11. Zeng W, Lei S (2003) Efficient frequency domain selective scrambling of digital video. IEEE Trans Multimedia 5(1):118–129
12. Bhargava B, Shi C, Wang S-Y (2004) MPEG video encryption algorithms. Multimedia Tools Appl 24(1):57–79
13. Lian S, Sun J, Wang J et al (2007) A chaotic stream cipher and the usage in video protection. Chaos Solitons Fractals 34(3):851–859
14. Zeng W, Lei S (2003) Efficient frequency domain selective scrambling of digital video. IEEE Trans Multimedia 5(1):118–129
15. Magli E, Grangetto M, Olmo G (2011) Transparent encryption techniques for H.264/AVC and H.264/SVC compressed video. Signal Process 91(5):1103–1114
16. Moffat A, Neal RM, Witten IH (1998) Arithmetic coding revisited. ACM Trans Inf Syst (TOIS) 16(3):256–294
17. Ahn J, Shim HJ, Jeon B et al (2004) Digital video scrambling method using intra prediction mode. In: Pacific-Rim conference on multimedia. Springer, Berlin, Heidelberg, pp 386–393
18. Angelides MC, Agius H (eds) (2010) The handbook of MPEG applications: standards in practice. Wiley
19. Lian S, Liu Z, Ren Z et al (2005) Selective video encryption based on advanced video coding. In: Pacific-Rim conference on multimedia. Springer, Berlin, Heidelberg, pp 281–290
20. Menezes AJ, Katz J, Van Oorschot PC et al (1996) Handbook of applied cryptography. CRC Press
21. Stutz T, Uhl A (2011) A survey of H.264 AVC/SVC encryption. IEEE Trans Circ Syst Video Technol 22(3):325–339

22. Farajallah M (2015) Chaos-based crypto and joint cryptocompression systems for images and videos. Thèse de doctorat. Université de Nantes
23. Shahid Z, Chaumont M, Puech W (2011) Fast protection of H.264/AVC by selective encryption of CAVLC and CABAC for I and P frames. IEEE Trans Circ Syst Video Technol 21(5):565–576
24. Asghar MN, Ghanbari M, Reed MJ (2012) Sufficient encryption with codewords and bin-strings of H.264/SVC. In: 2012 IEEE 11th international conference on trust, security and privacy in computing and communications. IEEE, pp 443–450
25. Dubois L, Puech W, Blanc-Talon J (2011) Fast protection of H.264, AVC by reduced selective encryption of CAVLC. In: 2011 19th European signal processing conference. IEEE, pp 2185–2189
26. Yeung S-KA, Zhu S, Zeng B (2009) Partial video encryption based on alternating transforms. IEEE Signal Process Lett 16(10):893–896
27. Yeung, S-KA, Zhu S, Zeng B (2011) Design of new unitary transforms for perceptual video encryption. IEEE Trans Circ Syst Video Technol 21(9):1341–1345
28. Van Wallendael G, De Cock J, Van Leuven S et al (2013) Format-compliant encryption techniques for high efficiency video coding. In: 2013 IEEE international conference on image processing. IEEE, pp 4583–4587
29. Van Wallendael G, Boho A, De Cock J et al (2013) Encryption for high efficiency video coding with video adaptation capabilities. IEEE Trans Consumer Electron 59(3):634–642
30. Shahid Z, Puech W (2013) Visual protection of HEVC video by selective encryption of CABAC binstrings. IEEE Trans. Multimedia 16(1):24–36
31. Dworkin M (2004) Recommendation for block cipher modes of operation: the CCM mode for authentication and confidentiality. National Institute of Standards and Technology
32. Sallam AI, Faragallah OS, El-Rabaie E-SM (2017) HEVC selective encryption using RC6 block cipher technique. IEEE Trans Multimedia 20(7):1636–1644
33. Sallam AI, El-Rabaie E-SM, Faragallah OS (2018) Efficient HEVC selective stream encryption using chaotic logistic map. Multimedia Syst 24(4):419–437
34. Lui O-Y, Wong K-W (2013) Chaos-based selective encryption for H.264/AVC. J Syst Softw 86(12):3183–3192
35. Altaf M, Ahmad A, Khan FA et al (2018) Computationally efficient selective video encryption with chaos based block cipher. Multimedia Tools Appl 77(21):27981–27995
36. Ganeshkumar D, Suresh A, Manigandan K et al (2019) A new one round video encryption scheme based on 1D chaotic maps. In: 2019 5th international conference on advanced computing & communication systems (ICACCS). IEEE, pp 439–444
37. Chandrasekaran J, Thiruvengadam SJ (2015) Ensemble of Chaotic and Naive approaches for performance enhancement in video encryption. Sci World J 2015
38. Peng F, Li H, Long M (2015) An Effective Selective Encryption Scheme for HEVC based on Rossler Chaotic system. In: Proceedings of International Symposium Nonlinear Theory its Application, pp 1–4
39. NIST, Information Technology Laboratory, Computer Security Resource Center. Dernier mise à jour le: 22/06/2020. Random Bit Generation, NIST SP 800-22: Download Documentation and Software. Disponible sur: https://csrc.nist.gov/projects/random-bit-generation/documentation-and-software [consulté le: 10/04/2020]

Neural Network Image Restoration Techniques

Mingyong Jiang, Ningbo Guo, Qun Wei, and Hang Chen

Abstract Due to the effections of environment and human factor, images always contains degradation phenomenon caused by image blur and noise which reduces the image quality. So, focus on the topic of the improvement of image, this dissertation pays much attention on the study of image degradation model, image restoration and evaluation method. Firstly, the image degradation model is given. Then, a few evaluation algorithms of restoration are given. At last, three Hopfield Neural Network restoration algorithm based on Laplace operator, sub-optimal algorithm and Harmonic Model are investigated. Simulation results validate the efficiency of the algorithms.

1 Introduction

Encrypted images are often distorted and blurred during transmission and use because of human attacks or noise contamination [1]. An important problem to be solved before the image is how to recover the original encrypted image from the degraded image. It is needed to establish a mathematical model of image degradation by using a priori knowledge of image degradation, and reverse the image degradation process to restore the original information of the image as much as possible. There are many reasons for the degradation of encrypted images, which can be usually classified into deterministic and random factors. Deterministic is mainly a human attack on the image and randomness refers to noise contamination [1].

Image restoration has important scientific research value and practical significance. In the project of using images as data, image degradation caused by any factor will reduce the scientific and practical value, leads to economic loss [2]. Neural network image restoration techniques have been applied to improve the quality of

M. Jiang · N. Guo · Q. Wei · H. Chen (✉)
School of Space Information, Space Engineering University, Beijing 101416, China
e-mail: hitchenhang@foxmail.com

N. Guo
Academy of People's Armed Police, Beijing 101416, China

the images and effectively recover the useful information of the images. It has great practical significance and is widely in many fields as follows [3]:

(1) Astronomical image processing: Overcoming the degradation of ground-based astronomical images caused by atmospheric disturbances, overcoming image degradation caused by imperfect observation and imaging systems, suppressing noise introduced by various factors, improving the quality of camera and video obtained by spacecraft.

(2) Optical remote sensing image processing: overcoming the degradation caused by atmospheric disturbance, overcoming the image degradation caused by the imperfect camera or imaging system and the movement of the remote sensing platform relative to the ground, suppressing noise and thin clouds.

(3) Synthetic Aperture Radar Remote Sensing Image processing: improve image resolution, suppress noise, suppress ghost images, suppress smearing, local geometric correction.

(4) Medical image processing: Suppress the noise of various medical imaging systems and improve the resolution of medical images.

(5) Material science image processing: improve the resolution of optical microscopy images, electron microscopy images, industrial X-CT and diffraction images. Kikuchi line enhancement and reconstruct molecular structure from X-ray diffraction images.

(6) Historical and Humanistic Photo Restoration.

(7) Surveillance video Restoration.

(8) Early Film Renewal.

(9) Video and multimedia image processing: overcome block pattern interference in decoded images due to high data compression and chunking transformations (e.g. DCT), overcome image degradation introduced by transcription systems.

(10) Scanned Document Processing: Overcome the interference of mesh in scanned document images, change halftone of document images to be continuous for improved visual effects.

2 Description of Image Restoration

Factors lead to image degradation can be summarized into ambiguity and noise, the former can be generally described by a ambiguity function and the latter can be described by a stochastic process. The process can be represented using a ambiguity function $h(x, y)$, which is a spatial description of the image energy distribution. Image noise is an unpredictable random error, which can only be analyzed statistically and described by a probability density function and a probability distribution function.

However, the exact distribution of image noise is impossible to obtain in practical applications and usually described using numerical characteristics. In most cases, image noise can be assumed to obey a Gaussian distribution. Which can be divided into additive and multiplicative noise according to the way it acts. The image is assumed as a grayscale image, represented by a two-dimensional function $f(x, y)$ and

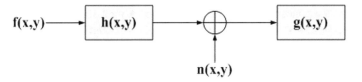

Fig. 1 Image degradation model

the noise n(x, y) can be regarded as interference with the grayscale. The degraded image contaminated by noise is assumed to be g(x, y). For additive noise, such as amplifier noise, g(x, y) = f(x, y) + n(x, y). For multiplicative noise, Such as quantum noise and particle noise, g(x, y) = f(x, y)[1 + n(x, y)]. The calculation of the image degradation model with multiplicative noise is complicated. When the noise variation is small, the multiplicative noise is generally approximated as additive noise to simplify the model and the image and noise are assumed to be statistically independent [1]. Based on the above analysis, the image degradation model can be represented as shown in Fig. 1.

If the degradation process is a linear shift-invariant process, the degraded image will be represented in the spatial domain as

$$g(x, y) = f(x, y) * h(x, y) + n(x, y) \tag{1}$$

where * is the convolution operation and h(x, y) is the spatial domain representation of the ambiguity function. The degenerate image can be represented in the frequency domain as

$$G(u, v) = F(u, v) \cdot H(u, v) + N(u, v) \tag{2}$$

where G(u, v), F(u, v), H(u, v) and N(u, v) are the Fourier transforms of g(x, y), f(x, y), h(x, y) and n(x, y), respectively.

When the degradation model is established, the purpose of image restoration is to obtain an approximate solution of the original image f with known a priori information. From the above degradation model, it can be seen that image degradation is a positive problem and image restoration is an inverse problem. According to the operator theory, we can find a transformation T between f and g, where the function f belongs to the space H1 and the function g belongs to the space H2 [1].

$$T\{f\} = g \tag{3}$$

The inverse conversion of image restoration can be written as

$$T^{-1}\{g\} = f \tag{4}$$

The image restoration is the study of the existence and uniqueness of the inverse permutation in mathematical meanings. In 1923, French mathematician Hadamard put forward the concept of well-being problems. A well-state problem satisfies the following conditions: (1) the solution of the problem exists, (2) the solution is unique, (3) the only solution of the problem is stable. If one of the three conditions is not met, the problem is pathological or ill-posed. Theoretically, the degenerate equation is the Fredholm integral equation of the first kind and its solution that depends on the data discontinuously is an ill-conditioned problem. When the observation data is disturbed, the solution space is too large. If the solution is discontinuously depends on the observation data, a small change may lead to a large change in the solution, which can be expressed as [1]

$$T^{-1}\{g + \varepsilon\} = f + \delta \tag{5}$$

where ε is an arbitrarily small value and δ is the disturbance caused by ε. It is an ill-conditioned problem when $\delta \gg \varepsilon$.

Because the true solution or approximate solution of the equation is in the solution space, additional restrictions are introduced to define a compact set containing the true solution and find the true solution in the intersection of the original solution space and this compact set. On the other hand, we can add a small additional term to the Fredholm equation of the first kind to make it a Fredholm equation of the second kind. Because the solution is stable in small changes and be close to the true solution of the original problem [1].

Therefore, in order to obtain useful results, some approximate method must be sought to modify the problem so that the modified problem becomes a good state. At the same time, the solution of the revised problem must be very close to the true solution of the original problem. We can use the prior knowledge to make the solution continuously and meaningfully. The important thing is the selection of additional items. Due to the diversity of physical problems and the differences in people's understanding and utilization of prior knowledge, different problems and different researchers will get different methods. The commonly used methods in signal restoration are as follows [1, 3]: 1. modify the concept of the problem. Such as turning the original problem into a minimization problem, turning the solution problem into an iterative or filtering process and modifying the solution using additional constraints during the iteration and filtering process, turning the solution problem into a projection iteration problem and turning the solution problem into a model estimation problem. 2. Restrict data. Such as generalized inverse and singular value decomposition are used to suppress the influence of data errors, the high-frequency components of the solution are estimated and truncated, non-linear filtering and projection are used to eliminate unreasonable data. 3. Modify the space or topology. Restrict the domain and construct a reasonable limit for the solution, making it belong to a compact set. 4. Modify the operator. For example, use regularization operators. 5. Use statistical estimation methods. Establish a statistical model of the solution and use the statistical model as a priori knowledge to turn the original problem into a statistical estimation or parameter optimization problem.

3 Evaluation of Image Restoration

It is necessary to evaluate the image quality for evaluating the performance of the image restoration in image restoration. Image quality evaluation includes subjective evaluation and objective evaluation. The subjective evaluation is a method that uses human eyes to evaluate images. The evaluation results are more accurate and reliable, but the results of subjective evaluation are only people's qualitative understanding of image quality and different people may get different results.

There are ideal images, degraded images and restored images in image restoration, so it is generally a reference objective evaluation method to use the joint measurement of the three images. Evaluation indicators in image restoration used commonly include Mean Square Error (MSE), Normalized Mean Square Error (NMSE), Signal to Noise Ratio (SNR), and Peak SNR Ratio (Peak Signal to Noise Ratio—PSNR) and Improved Signal to Noise Ratio (ISNR). In image restoration, we set the original image as f, the degraded image as g, and the restored image as F. F(i, j), f(i, j) and g(i, j) are the restored image, original image and The gray value of the degraded image at the coordinates (i, j). The gray value ranges from 0 to 255, and M and N are the size of the image. Evaluation indicators can be calculated by the following expression.

$$MSE = \frac{1}{MN}\|F - f\|^2 = \frac{\sum\limits_{i=1}^{M}\sum\limits_{j=1}^{N}\left[F(i,j) - f(i,j)\right]^2}{MN} \tag{6}$$

$$NMSE = \frac{\|F - f\|^2}{\|f\|^2} = \frac{\sum\limits_{i=1}^{M}\sum\limits_{j=1}^{N}\left[F(i,j) - f(i,j)\right]^2}{\sum\limits_{i=1}^{M}\sum\limits_{j=1}^{N}\left[f(i,j)\right]^2} \tag{7}$$

$$SNR = 10\log_{10}\left(\frac{\|f\|^2}{\|F - f\|^2}\right) = 10\log_{10}\left[\frac{\sum\limits_{i=1}^{M}\sum\limits_{j=1}^{N}\left[f(i,j)\right]^2}{\sum\limits_{i=1}^{M}\sum\limits_{j=1}^{N}\left[F(i,j) - f(i,j)\right]^2}\right] \tag{8}$$

$$PSNR = 10\log_{10}\frac{255^2}{MSE} = 10\log_{10}\left[\frac{255^2 \times MN}{\sum\limits_{i=1}^{M}\sum\limits_{j=1}^{N}\left[F(i,j) - f(i,j)\right]^2}\right] \tag{9}$$

$$ISNR = 10\log_{10}\frac{\|g - f\|}{\|F - f\|^2} = 10\log_{10}\left[\frac{\sum\limits_{i=1}^{M}\sum\limits_{j=1}^{N}\left[g(i,j) - f(i,j)\right]^2}{\sum\limits_{i=1}^{M}\sum\limits_{j=1}^{N}\left[F(i,j) - f(i,j)\right]^2}\right] \tag{10}$$

Among the above five evaluation indicators, MSE and NMSE represent the deviation between the restored image and the original image. Their values are smaller, the image quality will be better. The opposite is true for SNR, PSNR and ISNR. PSNR is commonly used in image restoration to indicate the degree of the degraded image, and ISNR indicates the PSNR of the restored image relative to the PSNR of the degraded image [4].

4 Neural Network Image Restoration

Neural networks are an important research component of artificial intelligence and widely employed in many fields because of superb fault tolerance, computational parallelism and nonlinear approximation, such as machine vision, pattern recognition, intelligent computing and system identification. Considering the network image restoration technique does not require the image to be generalized and stable, the network image restoration technique is widely used and usually divided into two categories one is to train the neural network with training samples composed of the original image and the degraded image. The trained neural network can be established by utilizing the nonlinear of the neural network to approximate the image degradation process. We can deal with the degradation Image to achieve the purpose of restoration with the trained neural network. The other is neural network iterative restoration. Hopfield Neural Network (Hopfield Neural Network—HNN) have been studied by a large number of scholars in neural network iterative restoration for solving optimization problems without generalized stationary assumptions [5].

Though Zhou et al. [5] developed the restoration of Hopfield neural network, there are some serious shortcomings by use this neural network to restore the image. Each pixel of the image needs several binary neurons to represent and the number is equal to the size of the pixel gray value, which leads to the large scale of the neural network. It is inefficient that the sign of the energy function of the neural network needs to be checked in each iterative update of the neuron state. In order to solve the above problem, Paik et al. proposed an improved Hopfield neural network model [6–8] and used to implement image restoration. Multi-valued neurons have been proposed to represent the pixel values of the image, which greatly reduces the size of the neural network. They also proposed a variety of mapping methods between images and neural networks to prove the convergence of various update rules. Yi Sun [6–8] also studied the Hopfield neural network restoration of various update rules. According to their investigation, Paik's method has a better effect in restoring severely blurred images without noise. The neuron state of the neural network used by Paik is discontinuous. Therefore, Lei Wang et al. [9] and YuBin Han et al. [10] proposed a Hopfield neural network image restoration method with continuous state changes, which overcomes the shortcomings of the step change of neuron state. It can be [11] showed that the Paik algorithm recovered ambiguity images better in the absence of noise. The Hopfield neural network image restoration method [12, 13] with continuous state changes has been proposed to overcomes the disadvantage of step changes in the neuron states of the Paik algorithm.

On the other hand, a number of researches have been presented to improve the performance of neural network recovery by adaptive regularization. Based on the statistical characteristics of the local variance of the image, Perry et al. [14] adopted different regularization parameters in the edge area and the non-edge area to realize the adaptive regularization neural network restoration. Yang et al. [15] have developed a neural network by vision principle to detect the edges of a blurred image and used different regularization parameters. EHE (Eliminating Highest Error) has been proposed to improve the image restoration of Hopfield neural network [16], which can better solve the inverse degradation process and smoothing Contradiction. Qian Wei et al. [17] Modified the regularization parameters according to the statistical properties of the brightness variance of the image. Hau-san Wong et al. [18] proposed a stochastic gradient descent algorithm based on regional spatial characteristics and a neural network with weights satisfying specific functional forms to optimize regional regularization parameters. Mohsin Bilal et al. [19] used BP neural network to determine whether the image field contains edges and have selected different regularization parameters to achieve adaptive restoration. Fuzzy quasi-range edge detector have been used to achieve adaptive regularization [20]. Regularization parameters [21, 22] have been selected according to the local statistical characteristics of the image to achieve adaptive restoration. Clustering analysis [23] have been used on the non-zero elements of the co-occurrence matrix and modified the regularization parameters according to the intensity of the details obtained from the analysis.

4.1 Energy Functions

Linear shift invariant image degradation can be expressed as

$$y = h * x + n \tag{11}$$

where x and y are the original image and the degraded image respectively, h is the ambiguity function, n is the additive noise, and * is the convolution symbol. It is usually expressed in the form of a vector matrix

$$Y = HX + N \tag{12}$$

where X, Y, and N are the vector forms of the original image, degraded image, and additive noise respectively. H and N respectively represent the two main factors of ambiguity and noise that cause image degradation. They are determined by

$$X = \begin{bmatrix} X_1 \\ X_2 \\ \cdots \\ X_M \end{bmatrix} \tag{13}$$

where

$$X_i = \begin{bmatrix} x_{(i-1)\times N+1} \\ x_{(i-1)\times N+2} \\ \vdots \\ x_{i\times N} \end{bmatrix} \tag{14}$$

Y and N have the same form as X.

h can be written as a convolutional form of a small window, considering the neighborhood range of the image, the ambiguity function can be expressed as

$$h(k, l) = \begin{cases} \frac{1}{2} & k = 0, \ l = 0 \\ \frac{1}{16} & \text{others} \end{cases} \tag{15}$$

the ambiguity matrix H is a block Toeplitz matrix or a block-circulant matrix when the image has periodic boundaries. The block-circulant matrix can be expressed as

$$H = \begin{bmatrix} H_0 & H_1 & O & \cdots & O & H_1 \\ H_1 & H_0 & H_1 & \cdots & O & O \\ \vdots & \vdots & \vdots & \cdots & \vdots & \vdots \\ H_1 & O & O & \cdots & H_1 & H_0 \end{bmatrix} \tag{16}$$

where

$$H_0 = \begin{bmatrix} \frac{1}{2} & \frac{1}{16} & 0 & \cdots & 0 & \frac{1}{16} \\ \frac{1}{16} & \frac{1}{2} & \frac{1}{16} & \cdots & 0 & 0 \\ \vdots & \vdots & \vdots & \cdots & \vdots & \vdots \\ \frac{1}{16} & 0 & 0 & \cdots & \frac{1}{16} & \frac{1}{2} \end{bmatrix} \quad H_1 = \begin{bmatrix} \frac{1}{16} & \frac{1}{16} & 0 & \cdots & 0 & \frac{1}{16} \\ \frac{1}{16} & \frac{1}{16} & \frac{1}{16} & \cdots & 0 & 0 \\ \vdots & \vdots & \vdots & \cdots & \vdots & \vdots \\ \frac{1}{16} & 0 & 0 & \cdots & \frac{1}{16} & \frac{1}{16} \end{bmatrix} \tag{17}$$

The size of H is MN × MN, the size of H0 and H1 is M × N, and O is a zero matrix of the same size as H0 and H1 [5].

According to Lagrange multiplier approach, Eq. 12 solves the problem of X optimal solution, which is consistent with the minimization of the cost function shown in Eq. 18.

$$J\left(\hat{X}\right) = \frac{1}{2}\left\| Y - H\hat{X} \right\|^2 + \frac{1}{2}\lambda \left\| D\hat{X} \right\|^2 \tag{18}$$

where $\|Z\|$ is norm of Z and $\lambda > 0$ is the Lagrange multiplier, also called the regularization parameter. If λ is larger, its noise suppression ability is stronger, but it will lead to image smoothing and even ringing effect. λ should be chosen to take into account both noise suppression and image detail protection [5].

In traditional neural network image restoration, D is the regular matrix that generated using the Laplace operator and can be written as

$$\nabla = \frac{\partial^2}{\partial x^2} + \frac{\partial^2}{\partial y^2} \tag{19}$$

The convolution window represented by the Laplace operator can be written as

$$\frac{1}{6}\begin{bmatrix} 1 & 4 & 1 \\ 4 & -20 & 4 \\ 1 & 4 & 1 \end{bmatrix} \tag{20}$$

D is also a block-circulant matrix with dimensions as shown in Eqs. 16 and 18 can be expanded as

$$J\left(\hat{X}\right) = \tfrac{1}{2}\hat{X}^T\left(H^TH + \lambda D^TD\right)\hat{X} - \left(H^TY\right)\hat{X} + \tfrac{1}{2}Y^TY \tag{21}$$

The energy function of a neural network [7] can be defined as

$$E(v) = -\tfrac{1}{2}v^TTv - Bv \tag{22}$$

where v is the state of the neuron, T is the connection power matrix, and B is the neuron bias matrix. Comparing Eqs. 21 and 22 and removing the constant terms, we can derive the mapping relationship as

$$v = \hat{X}, T = -\left(H^TH + \lambda D^TD\right), B = H^TY \tag{23}$$

The problem of minimizing the cost function shown in Eq. 18 is transformed into the problem of minimizing the energy function of the Hopfield neural network shown in Eq. 22, which can be calculated by the state evolution of the neural network.

4.1.1   Algorithm of Zhou

The neural network of Zhou's algorithm is shown in Fig. 2, where xi represents the ith pixel of the image, Blocki represents the mapping scheme of pixel i, si represents the order of switches in updating, and different combinations of switches represent different updating methods, which are divided into fully parallel updating method (all si are turned on), serial updating method (si are turned on sequentially in order) and partially parallel updating [7].

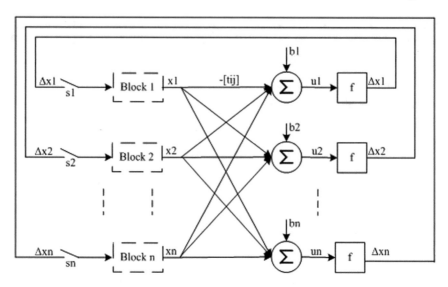

Fig. 2 Hopfield neural network used by Zhou's algorithm

The Block in Fig. 2 can be replaced by the module in Fig. 3, (a) for simple summation and (b) for binary summation. The simple summation uses 255 (for a grayscale image with a gray level of 255) variables v to represent individual pixels whose values take only 0 or 1, and the sum of all v is the current pixel value. Binary summation uses eight v to represent a single pixel and the current pixel value can be obtained by adding up according to different weights. Using the pixel values of the degraded image as the initial values of the neural network, Zhou uses a serial update and the process can be written as

Update Rule 1 [7]: (Simple summation-serial update)

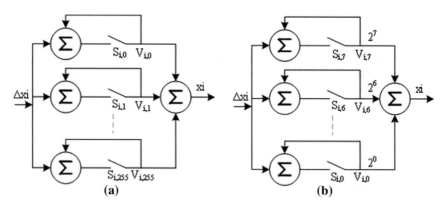

Fig. 3 Two mappings of pixel values in Zhou's algorithm: (**a**) simple summation scheme, (**b**) binary summation scheme

For i = 1,…,n do

For k = 1,…,255 do

$$u_i = b_i - \sum_{j=1}^{n} x_j \tag{24}$$

$$\Delta x_i = d(u_i) = \begin{cases} -1 & u_i < 0 \\ 0 & u_i = 0 \\ 1 & u_i > 0 \end{cases} \tag{25}$$

$$\Delta f = \tfrac{1}{2} t_{ii} (\Delta x_i)^2 - u_i \Delta x_i \tag{26}$$

If $\Delta f < 0$ then $v_{i,k} = v_{i,k} + \Delta x_i$ (27)

$$x_i = \sum_{k=0}^{255} v_{i,k} \tag{28}$$

End
End

where $V_{i,k}$ is the k-th of 255 neurons representing the i-th pixel, and Δf represents the change in the energy of the neural network after the update. If the energy decreases, the updated value is used, and if the energy increases, the neuron is not updated. The energy detection in each iteration is indispensable because the algorithm does not guarantee that the energy is monotonically decreasing. The mapping in binary summation can be expressed as

$$x_i = \sum_{k=0}^{7} 2^k v_{i,k} \tag{29}$$

Update Rule 2: (Parallel method)
In the parallel update method, all neurons are updated based on the neuron state at the previous moment. xt can be assumed to represent the neuron state at the current moment and xt + 1 represents the updated neuron state, the flow chart can be expressed as

For i = 1,…,n

For k = 1,…,255

$$u_i = b_i - \sum_{j=1}^{n} x_j^t \tag{30}$$

$$\Delta x_i = d(u_i) = \begin{cases} -1 & u_i < 0 \\ 0 & u_i = 0 \\ 1 & u_i > 0 \end{cases} \tag{31}$$

$$\Delta f = \tfrac{1}{2} t_{ii}(\Delta x_i)^2 - u_i \Delta x_i \tag{32}$$

$$\text{If } \Delta f < 0 \text{ then } v_{i,k}^{t+1} = v_{i,k}^t + \Delta x_i \tag{33}$$

$$x_i^{t+1} = \sum_{k=0}^{255} v_{i,k}^t \tag{34}$$

End
End

Update Rule 3: (Partial parallelism)
While update is not completed
Determine the set of neurons A that need to be updated, $A = \{A(i) = x, x \in 1,\ldots,$
n and non-repeating, $i = 1,\ldots, L\}$. Where L is the number of neurons that need to be
updated simultaneously and elements in A represent the L current neurons that need
to be updated.

For $i = 1,\ldots,n$

   For $k = 1,\ldots,255$

$$u_{A(i)} = b_{A(i)} - \sum_{j=1}^{n} x_j^t \tag{35}$$

$$\Delta x_{A(i)} = d\left(u_{A(i)}\right) = \begin{cases} -1 & u_{A(i)} < 0 \\ 0 & u_{A(i)} = 0 \\ 1 & u_{A(i)} > 0 \end{cases} \tag{36}$$

$$\Delta f = \tfrac{1}{2} t_{A(i)A(i)}\left(\Delta x_{A(i)}\right)^2 - u_{A(i)} \Delta x_{A(i)} \tag{37}$$

$$\text{If } \Delta f < 0 \text{ then } v_{A(i),k}^{t+1} = v_{A(i),k}^t + \Delta x_{A(i)} \tag{38}$$

$$x_{A(i)}^{t+1} = \sum_{k=0}^{255} v_{A(i),k}^t \tag{39}$$

End
End

Zhou's algorithm checks the decay of the energy function at each iteration and will update the neural network if the update makes the energy function decrease.

4.1.2 Algorithm for Paik

Based on Zhou's algorithm, Paik uses multi-valued neurons to improve the mapping scheme between image and neuron states, greatly reducing the size of the neural network and simplifying the neural network image restoration. The neural network used by Paik's algorithm is shown in Fig. 4, where xi is a variable that takes values from 0 to 255.

Its update rules are

For $i = 1,\ldots,n$

$$u_i = b_i - \sum_{j=1}^{n} t_{ij}x_j(t) = e_i^T (b - Tx(t)) \qquad (40)$$

$$\Delta x_i = d_i(u_i) = \begin{cases} -1 & u_i < -\theta_i \\ 0 & -\theta_i < u_i < \theta_i \\ 1 & u_i > \theta_i \end{cases} \qquad (41)$$

$$\text{where } \theta_i = \tfrac{1}{2}t_{ii} > 0 \qquad (42)$$

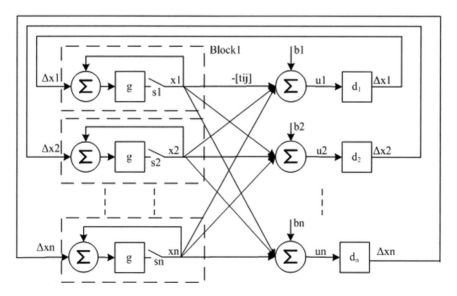

Fig. 4 Hopfield neural network used by Paik's algorithm

$$x_i(t+1) = g(x_i(t) + \Delta x_i), \qquad i = 1, ..., n \tag{43}$$

End

in which

$$g(v) = \begin{cases} 255 & v > 255 \\ v & 0 \le x \le 255 \\ 0 & x < 0 \end{cases} \tag{44}$$

where ei denotes the i-th unit vector, the initial values of the neural network use the pixel values of the degraded image. Compared with Zhou's algorithm, the main difference in the update rule used by Paik's algorithm is that the algorithm uses negative connection coefficients, which ensure the decreasing energy function, thus eliminating the need to check the decay of the energy function during the iteration.

4.1.3 Continuous Hopfield Neural Network Algorithm

The structure of the improved Continuous Hopfield Neural Network (CHNN) is shown in Fig. 5.

The update rule can be written as

For i = 1,...,n

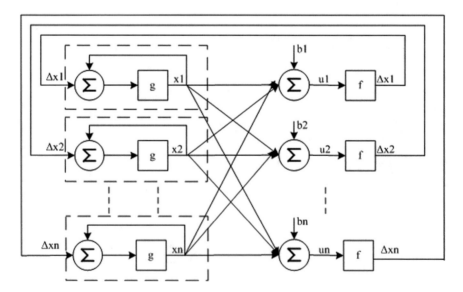

Fig. 5 Continuous Hopfield neural network algorithm

$$u_i = b_i - \sum_{j=1}^{MN} t_{i,j} \hat{x}_j \qquad (45)$$

$$\Delta x_i = f(u_i) \qquad (46)$$

$$g(x) = \begin{cases} 0.5 & x > 0.5 \\ x & -0.5 \le x \le 0.5 \\ -0.5 & x < -0.5 \end{cases} \qquad (47)$$

$$x_i(t+1) = g(x_i(t) + \Delta x_i) \qquad (48)$$

End

Where f(ui) can be in the following two forms [31].

$$f(u_i) = ru_i, \quad r > 0 \qquad (49)$$

$$f(u_i) = \frac{2}{1 + e^{-ru_i}} - 1, \quad r > 0 \qquad (50)$$

where r is the adjustment parameter, the algorithm converges faster when r increases, but too large r can cause the algorithm to be unstable, the stability conditions of the algorithm have been studied [13].

Theorem 1 When f takes Eq. 49, the convergence condition for the recovery of the serially updated Hopfield neural network is $0 < r < \frac{2}{\max(t_{ii})}$.

Proof When the state of the ith neuron changes from xi to xi + Δxi, the amount of change in the energy function of the neural network is

$$\Delta E = 0.5 t_{ii} (\Delta x_i)^2 - u_i \Delta x_i \qquad (51)$$

if ΔE < 0, we can obtain

$$|\Delta x_i| < \frac{2|u_i|}{t_{ii}} \qquad (52)$$

Substituting Eq. 49 into Eq. 52, we can obtain that the serially updated Hopfield neural network energy function is decaying when $0 < r < \frac{2}{\max(t_{ii})} < \frac{2}{t_{ii}}$.

Theorem 2 When f takes Eq. 49, the convergence condition for parallel update Hopfield neural network restoration is $0 < r < \frac{2}{\|T\|_2}$.

Proof Let $l(t) = \{1,2,...,n\}$

$$s(t) = x(t+1) - x(t) = \sum_{i \in l(t)} \Delta x_i e_i = \sum_{i=1}^{n} \Delta x_i e_i \tag{53}$$

$$E(x(t+1)) = E(x(t) + s(t)) = E(x(t)) + s(t)^T (Tx - b) + \frac{1}{2} s(t)^T s(t)$$

$$= E(x(t)) + \sum_{i=1}^{n} \Delta x_i e_i^T (Tx - b) + \frac{1}{2} \left(\sum_{i=1}^{n} \Delta x_i e_i \right)^T T \sum_{i=1}^{n} \Delta x_i e_i$$

$$= E(x(t)) - \sum_{i=1}^{n} \Delta x_i u_i + \frac{1}{2} \left(\sum_{i=1}^{n} \Delta x_i^2 \right) z^T T z$$

$$\leq E(x(t)) - \left[\sum_{i=1}^{n} \Delta x_i u_i - \frac{1}{2} \|T\|_2 \sum_{i=1}^{n} \Delta x_i^2 \right]$$

$$= E(x(t)) - \sum_{i=1}^{n} \left(\Delta x_i u_i - \frac{1}{2} \|T\|_2 \Delta x_i^2 \right) \tag{54}$$

Thus we can obtain

$$\Delta E = E(x(t+1)) - E(x(t)) \leq -\sum_{i=1}^{n} \left(\Delta x_i u_i - \frac{1}{2} \|T\|_2 \Delta x_i^2 \right) \tag{55}$$

From the quadratic spectrum theorem we can get

$$|\Delta x_i| < \frac{2|u_i|}{\|T\|_2} \tag{56}$$

The energy function of the Hopfield neural network updated in parallel is decaying when $0 < r < \frac{2}{\|T\|_2}$.

Theorem 3 When f takes Eq. 50, the convergence condition of the serial update Hopfield neural network restoration is $0 < r < \frac{4}{\max(t_{ii})}$.

Proof $\Delta E < 0$ when $|\Delta x_i| < \frac{2|u_i|}{t_{ii}}$ and Substituting Eq. 50, we can get $\left| \frac{2}{1+e^{-ru_i}} - 1 \right| < \frac{2|u_i|}{t_{ii}}$. Where Δx_i and u_i have the same sign, Let $H(u_i) = f(u_i) - \frac{2u_i}{t_{ii}}$ and $s(x) = \frac{1}{1+e^{-x}}$.

$$\frac{dH(u_i)}{du_i} = -2rs^2(ru_i) + 2rs(ru_i) - \frac{2}{t_{ii}} \tag{57}$$

The above equation is quadratic in s(rui) and the discriminant is less than zero. The serially updated Hopfield neural network energy function is decaying when $r < \frac{4}{\max(t_{ii})} < \frac{4}{t_{ii}}$.

Theorem 4 When f takes Eq. 50, the convergence condition of parallel update Hopfield neural network restoration is $0 < r < \frac{4}{\|T\|_2}$.

Proof $\Delta E = E(x(t+1)) - E(x(t)) \leq -\sum_{i=1}^{n} (\Delta x_i u_i) - \frac{1}{2}\|T\|_2 \Delta x_i^2$, $\Delta E < 0$ when $r < \frac{4}{\|T\|_2}$ and the energy function of the Hopfield neural network updated in parallel is decaying.

4.1.4 Simulated Results

The simulation experiment is used to compare the performance of Zhou's algorithm, Paik's algorithm and continuous Hopfield algorithm. The performance of the algorithm is compared from the attenuation of the energy function, the improvement of the image PSNR and the image comparison.

As shown in Fig. 6, the performance of the algorithm is tested by intercepting the 64×64 image from the standard Lenna image and the standard Cameraman image. The image uses a Gaussian blur with a size of 3×3 and $\sigma = 10$, adds Gaussian noise with a mean value of zero and a normalized variance of 0.118. The improved signal-to-noise ratio curve, energy function curve and recovery time are used to evaluate the performance of the algorithm.

The results of the three algorithms to restore the Lenna image are shown in Fig. 7. Compared with degraded images, the three algorithms can greatly improve the quality of the image in terms of visual effects and the effect of noise suppression is more obvious. It can be seen from Table 1 that the improved signal-to-noise ratio of the restoration results of the three algorithms has been greatly improved. Among them, Zhou's algorithm has the smallest increase of 5.4191 dB, and the continuous Hopfield algorithm (CHNN) has the largest increase of 6.254 dB. In order to achieve the same

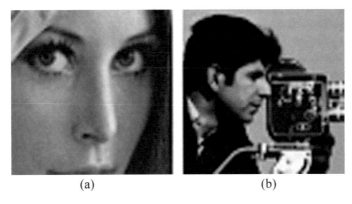

(a)   (b)

Fig. 6 The captured standard Lenna and Cameraman images: (**a**) the captured image of Lenna, (**b**) the captured image of Cameraman

recovery purpose, the time of Zhou's algorithm is much longer than Paik's algorithm and continuous Hopfield neural network algorithm, and it is almost 30 times longer than the latter two algorithms. Therefore, Zhou's algorithm is relatively inefficient. Paik's algorithm and continuous Hopfield algorithm adopt a more reasonable mapping method, and the algorithm performance has been greatly improved. This can also be seen from the results of the three algorithms to restore the Cameraman image as shown in Fig. 8 and Table 2.

Figure 9 shows the improvement of the signal-to-noise ratio as the number of iterations increases when the three algorithms restore the Lenna image. It can be seen that in terms of the number of iterations, Zhou's algorithm and CHNN's algorithm

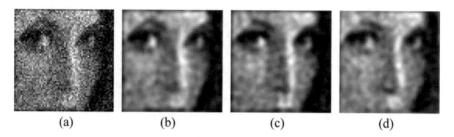

| (a) | (b) | (c) | (d) |

Fig. 7 Results of Lenna by Hopfield neural network algorithm based on Laplace operator: (**a**) degraded image, (**b**) result of Zhou, (**c**) result of Paik, (**d**) result of CHNN

Table 1 Comparison of Hopfield neural network algorithm based on Laplace operator (1)

Lenna	Zhou	Paik	CHNN
ISNR/dB	5.4191	5.8089	6.254
Time/s	5531	180	168

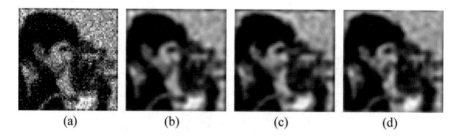

| (a) | (b) | (c) | (d) |

Fig. 8 Results of Cameraman by Hopfield neural network algorithm based on Laplace operator: (**a**) degraded image, (**b**) result of Zhou, (**c**) result of Paik, (**d**) result of CHNN

Table 2 Comparison of Hopfield neural network algorithm based on Laplace operator (2)

Cameraman	Zhou	Paik	CHNN
ISNR/dB	1.6396	1.5313	2.1486
Time/s	4319	183	162

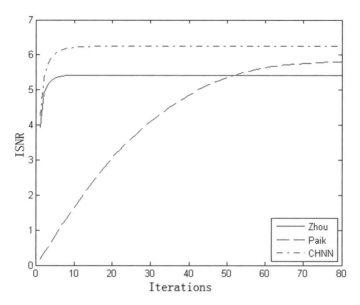

Fig. 9 The improved signal-to-noise ratio curve of the Hopfield neural network algorithm based on the Laplacian operator to restore the Lenna image

are the first to converge, Paik's algorithm converges relatively smoothly, but Zhou's algorithm takes a longer time for a single iteration. In terms of time, Zhou's algorithm Convergence is the slowest. The Paik algorithm and the CHNN algorithm cost the same amount of time and calculation in a single iteration. It can be concluded that the CHNN algorithm is better than Paik's algorithm in terms of convergence speed. A conclusion similar can be drawn in Fig. 10.

It can be seen from the simulation results that the three algorithms can achieve the restoration of degraded images and the image quality has been greatly improved. However, Zhou's algorithm converges the slowest, and it costs a huge amount of time and calculations. Paik's algorithm improves the inefficiency of Zhou's algorithm in image mapping and its time-consuming and computational complexity have been greatly improved. Based on the Paik algorithm, the CHNN algorithm further improves the convergence speed of the algorithm and is the best algorithm among the three algorithms. However, the restoration results of the three algorithms are relatively fuzzy. The reason is that the Laplacian operator is used, which only plays the role of image smoothing in the restoration process.

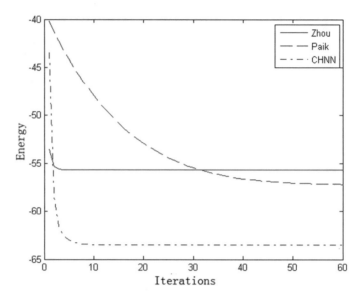

Fig. 10 The Hopfield neural network algorithm based on the Laplacian operator to restore the energy curve of the Lenna image

4.2 Suboptimal Algorithm of Continuous Hopfield Neural Network

The restoration method above costs huge resources. The ambiguity matrix H and the regular matrix D are both 65,536 × 65,536 double-precision numerical matrices when the image size is 256 × 256. It is found that the gray value of a certain pixel of the image is closely related to the surrounding pixels during the degradation process, the smaller the distance, the greater the impact; otherwise, the impact is smaller. The characteristic of the ambiguity matrix also proves this point. Most elements in the ambiguity matrix are zero, only a small part of the elements are non-zero, and the position of the non-zero element is exactly in the neighborhood of the pixel. The regular matrix D also has similar properties. Therefore, image restoration based on neighborhood pixel information is reasonable. Compared with restoration based on overall image information, restoration based on neighborhood information is not an optimal restoration and the result is suboptimal.

The CHNN suboptimal restoration algorithm based on Laplace operator is as follows

(1) Extend the original image under cyclic boundary conditions. The pixels (i, j) of the original image are selected sequentially, and the neighborhood sub-images of the pixels are determined in the expanded image. It can be seen in Fig. 11 that the regions R1 and R3 are the neighborhood sub-images of (i, j).

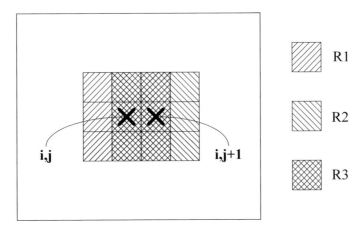

Fig. 11 Schematic diagram of neighborhood

(2) Calculate the ambiguity matrix H and the regular matrix D according to the point spread function, Laplacian operator and neighborhood sub-image.

(3) Calculate the connection weight matrix and bias vector according to the sub-image information, initialize the neural network.

(4) Use the CHNN algorithm to restore the sub-image of (i, j).

(5) Select the pixel value at position (i, j) in the restoration result as the final restoration result of pixel (i, j).

(6) Restore the next pixel (i, j + 1) until it is complete. The regions R2 and R3 are the neighborhood sub-images of (i, j + 1).

A partial image with a size of 64 × 64 is intercepted from the standard Lenna image as image 1, and a partial image with a size of 64 × 64 is intercepted as image 2 from the standard Cameraman image. The image is processed by Gaussian blur with a size of 3 × 3 and σ = 10, and Gaussian noise with a mean value of zero and a normalized variance of 0.078 is added. Under the parameter conditions of r = 0.1and λ = 0.5, the optimal restoration algorithm of CHNN based on Laplace operator and the sub-optimal CHNN algorithm of this section are used to restore the image. The peak signal to noise ratio (PSNR-Peak Signal to Noise Ratio) is used to measure the degree of image degradation, and the improved signal to noise ratio (ISNR-Improvement of SNR), the number of iterations and the restoration time are used to evaluate the performance of the algorithm. The results are shown in Table 3, Figs. 12 and 13.

Compared with the CHNN algorithm, the improved signal-to-noise ratio of the restoration result is slightly lower for the CHNN sub-optimal algorithm. As is shown in Table 3, the sub-optimal algorithm has a small loss of improved signal-to-noise ratio when restoring a partial Lenna image, when restoring a partial Cameraman image, the loss of signal-to-noise ratio is slightly larger. The number of sub-optimal restoration iterations is larger, but since the sub-optimal restoration is iteratively calculated within a small image range in the neighborhood, the restoration time is

Table 3 Comparison of CHNN algorithm and CHNN sub-optimal algorithm

		Reduced image PSNR	Restoration images PSNR	Restoration images ISNR/dB	Iteration time/s	time/s
1	CHNN	66.4428	71.1556	4.7128	181	230
	Sub-optimal	66.2595	70.9432	4.6837	460	190
2	CHNN	64.405	66.5064	2.1014	118	175
	Sub-optimal	64.2915	65.7712	1.4797	465	55

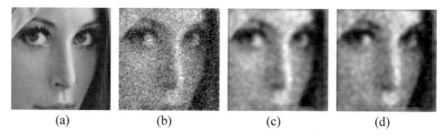

(a)   (b)   (c)   (d)

Fig. 12 Results of using the CHNN algorithm and the CHNN sub-optimal algorithm to restore the Lenna image: (**a**) original image, (**b**) degraded image, (**c**) CHNN algorithm, (**d**) CHNN sub-optimal algorithm

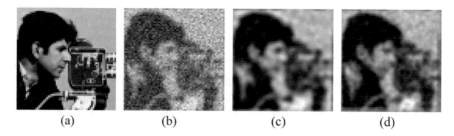

(a)   (b)   (c)   (d)

Fig. 13 Results of using the CHNN algorithm and the CHNN sub-optimal algorithm to restore the Cameraman image: (**a**) original image, (**b**) degraded image, (**c**) CHNN algorithm, (**d**) CHNN sub-optimal algorithm

not necessarily longer than the original algorithm. In the restoration of Lenna's partial image, the restoration time of the sub-optimal algorithm is slightly longer, and in the restoration of Cameraman's partial image, the restoration time of the sub-optimal algorithm is shorter than that of the original algorithm. Figures 12 and 13 show the restoration results of the CHNN algorithm and the CHNN suboptimal algorithm. The restoration of the sub-optimal algorithm is not limited by the image size and can realize the restoration of large-size images.

4.3   Continuous Hopfield Neural Network Suboptimal Restoration Algorithm Based on Harmonic Model

The Laplace operator in the traditional restoration algorithm only smooths the image without considering the edge information of the image. In the image restoration, it is easy to cause the image to be blurred while suppressing noise. In order to overcome this shortcoming, the harmonic model is applied to the Hopfield neural network, and the CHNN restoration algorithm based on the harmonic model is obtained [28].

4.3.1   CHNN Sub-Optimal Restoration Algorithm Based on Harmonic Model

We can choose a harmonic model to replace the Laplacian. Then the cost function becomes

$$J\left(\hat{X}\right) = \tfrac{1}{2}\left\|Y - H\hat{X}\right\|^2 + \tfrac{1}{2}\lambda\left\|\left(D_1\hat{X}\right)^2 + \left(D_2\hat{X}\right)^2\right\| \tag{58}$$

where D1 and D2 are the vertical gradient operator and the horizontal gradient operator, which are used to extract the edges of the image in the vertical and horizontal directions. The vertical and horizontal gradient operators can be expressed as

$$\frac{1}{4}\begin{bmatrix} -1 & 0 & 1 \\ -2 & 0 & 2 \\ -1 & 0 & 1 \end{bmatrix} \quad \frac{1}{4}\begin{bmatrix} -1 & -2 & -1 \\ 0 & 0 & 0 \\ 1 & 2 & 1 \end{bmatrix}$$

The expansion of Eq. 58 can be obtained as

$$J\left(\hat{X}\right) = \frac{1}{2}\sum_{p=1}^{MN}\left(y_p - \sum_{i=1}^{MN} h_{p,i}X_i\right)^2 + \frac{1}{2}\lambda\sum_{p=1}^{MN}\left(\left(\sum_{i=1}^{MN} D1_{p,i}X_i\right)^2\right.$$

$$\left. + \left(\sum_{i=1}^{MN} D2_{p,i}X_i\right)^2\right)$$

$$= \frac{1}{2}\sum_{i=1}^{MN}\sum_{j=1}^{MN}\left(\sum_{p=1}^{MN} h_{p,i}h_{p,j} + \lambda\sum_{p=1}^{MN} D1_{p,i}D1_{p,j} + \lambda\sum_{p=1}^{MN} D1_{p,i}D1_{p,j}\right)\hat{X}_i\hat{X}_j$$

$$- \sum_{i=1}^{MN}\left(\sum_{p=1}^{MN} y_p h_{p,i}\right)x_i + \frac{1}{2}\sum_{p=1}^{MN} y_p^2$$

$$= \frac{1}{2}\hat{X}^T\left(H^T H + \lambda D^T D\right)\hat{X} - \left(H^T Y\right)\hat{X} + \frac{1}{2}Y^T Y \tag{59}$$

The energy function of the neural network can be expressed as

$$E(v) = -\frac{1}{2}\sum_{i=1}^{MN}\sum_{j=1}^{MN} v_{j,i}T_{i,j}v_{i,j} - \sum_{i=1}^{MN} B_i v_i$$
$$= -\frac{1}{2}v^T Tv - Bv \tag{60}$$

Remove the constant term in Eq. 59 and we can get the following mapping relationship.

$$T = -\left(H^T H + \lambda D_1^T D_1 + \lambda D_2^T D_2\right), \qquad v = \hat{X}, \qquad B = H^T Y \tag{61}$$

The restoration process of the CHNN sub-optimal algorithm based on the reconciliation model can be written as follows

(1) Determine the size of the neighborhood sub-image in the sub-optimal algorithm according to the dimension of the point spread function. Generally, the size of the neighborhood is the same as the point spread function.
(2) Get the ambiguity matrix H according to the point spread function and the size of the neighborhood sub-image, get the regular matrices D1 and D2 according to the Sobel vertical gradient operator and the horizontal gradient operator.
(3) According to the size of the neighborhood, expand the original image under cyclic boundary conditions, and the expanded image contains the neighborhood of all pixels of the original image.

For k = 1: L (L is the total image size)

Determine the neighborhood sub-image Y of the k-th pixel.
Calculate the weight matrix T and the bias vector B of the neural network according to the neighborhood sub-image and Eq. 61.

Initialize the neural network with Y, X0 = Y.
While True
For i = 1: MN (M和N为邻域子图像的维数)

$$u_i = b_i - \sum_{j=1}^{MN} t_{i,j}\hat{x}_j \tag{62}$$

$$\Delta x_i = f(u_i) \tag{63}$$

$$g(x) = \begin{cases} 0.5 & x > 0.5 \\ x & -0.5 \le x \le 0.5 \\ -0.5 & x < -0.5 \end{cases} \tag{64}$$

$$x_i(t+1) = g(x_i(t) + \Delta x_i) \tag{65}$$

End

Calculate the network energy. When the network energy no longer changes or jumps between two values, the algorithm terminates.

End

Take the center pixel value of the restored domain sub-image as the corresponding pixel value of the final restoration result.

End

When the i-th update is completed, the change of the i-th neuron xi is Δxi.

$$\Delta E = -0.5t_{ii}(\Delta x_i)^2 - x_i\Delta x_i \tag{66}$$

ΔE is a quadratic function with the opening upward because tii < 0. ΔE is decreasing when r < −1/tii. This algorithm can ensure the convergence of the neural network energy function.

The standard Lenna image and the standard Cameraman image have been selected and a Gaussian blur with a size of 3 × 3 and σ = 10 can be used to process the image, which added Gaussian noise with a mean value of zero and a normalized variance of 0.118. Under the condition of r = 0.1 and λ = 0.5, the image is restored with the CHNN sub-optimal restoration algorithm based on the traditional Laplacian operator and the CHNN sub-optimal algorithm based on the harmonic model proposed in this section. Use the improved signal-to-noise ratio (ISNR-Improvement of SNR) to evaluate the performance of the algorithm.

It can be seen from Table 4 that the algorithm proposed is better than the traditional Laplacian-based CHNN suboptimal restoration algorithm. When using the algorithm to restore the Lenna standard image, the ISNR is slightly better than the traditional algorithm, when restoring the Cameraman standard image, the ISNR is greatly improved. It can also be seen from the restoration effect comparison graphs shown in Figs. 14 and 15 that the resolution of the restored image with the algorithm is better than that of the traditional image.

Table 4 Comparison of the suboptimal CHNN algorithm based on the Laplace operator and the reconciliation model		The Laplace operator ISNR/dB	This section ISNR/dB
	Lenna	6.2417	6.2704
	Cameraman	4.4124	5.1847

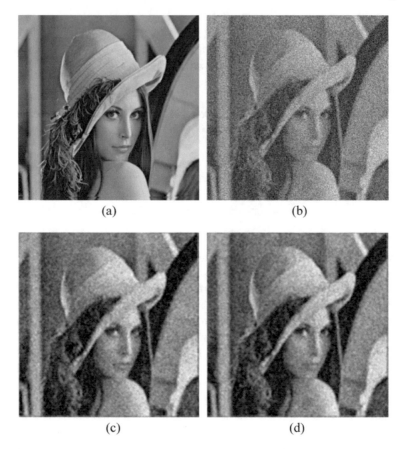

Fig. 14 Results of the CHNN suboptimal algorithm based on the reconciliation model and the recovery of Lenna by the algorithm based on the Laplace operator: (**a**) original image, (**b**) degraded image, (**c**) CHNN suboptimal algorithm based on Laplace operator, (**d**) results of the algorithm in this section

4.4 Adaptive CHNN Suboptimal Restoration Algorithm Based on Reconciliation Model

Both noise suppression and detail protection must be taken into account when select the regularization parameter λ in the surrogate function. The λ value should be increased in the flat area of the image to suppress noise as much as possible, while the λ value should be decreased in the edge area to focus on protecting edge details. Therefore, the edge detection operator is applied to process the image before image restoration, and the λ value is modified according to the result of edge detection to achieve suboptimal image restoration by adaptive regularized Hopfield neural network with edge detection. Taking the sobel operator as an example and the flow chart can be written as follow.

(a)   (b)

(c)   (d)

Fig. 15 Results of the CHNN suboptimal algorithm based on the reconciliation model and the algorithm based on the Laplace operator to recover Cameraman: (**a**) original image, (**b**) degraded image, (**c**) CHNN suboptimal algorithm based on Laplace operator, (**d**) results of the algorithm in this section

(1) Detect the edges of the image with the sobel operator.
(2) Modify the value of λ according to the detection result.
(3) Recover the image using the CHNN suboptimal algorithm based on the reconciliation model.

Taking standard Lenna graphs and Cameraman graphs as examples, the image is processed with Gaussian blur with a size of 3×3 and $\sigma = 10$, and Gaussian noise with a mean value of zero and a normalized variance of 0.078 is added. The restoration result of the Lenna graph is shown in Fig. 16. Compared with the degraded image, the sub-optimal restoration algorithm and the adaptive sub-optimal restoration algorithm achieve noise suppression. The adaptive algorithm preserves the edge information of the image better because it selects different regularization parameters in the flat and edge regions of the image, which can be concluded from the comparison of (c) and (d). The algorithm performance evaluation shown in Table 5 also shows that the adaptive algorithm is slightly better than the non-adaptive algorithm.

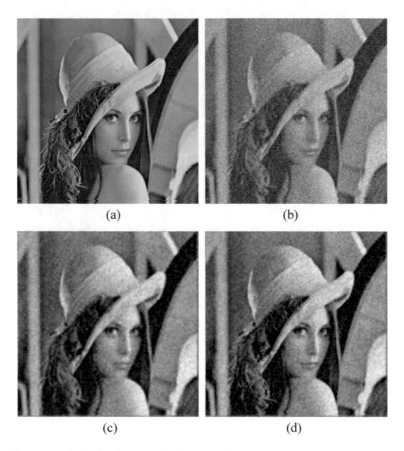

(a)   (b)

(c)   (d)

Fig. 16 Results of adaptive CHNN suboptimal algorithm and non-adaptive algorithm for recovering Lenna based on reconciliation model: (**a**) original image, (**b**) degraded image, (**c**) non-adaptive algorithm, (**d**) adaptive algorithm recovery result

Table 5 Comparison of adaptive CHNN suboptimal and non-adaptive algorithms based on the reconciliation model		Non-adaptive algorithms ISNR/dB	Adaptive algorithm ISNR/dB
	Lenna	4.3437	4.3822
	Cameraman	3.2701	3.6923

5 Conclusions

This article introduces the mathematical description of image restoration problem and the evaluation method of image restoration, studies three Hopfield neural network algorithms—Zhou's algorithm, Paik's algorithm and CHNN's algorithm. The Paik algorithm improves the image to neural network mapping based on Zhou's algorithm, which greatly reduces the size of the neural network and shortens the recovery time. The CHNN algorithm improves the disadvantage of discontinuous neural network state changes in the Paik algorithm and further improves the algorithm performance. In order to overcome the problem of occupying large resources in Hopfield neural network recovery, the suboptimal recovery algorithm of Hopfield neural network based on neighborhood information is proposed according to the finite support domain property of ambiguity function. Simulation results show that the algorithm can recover images of larger size with little loss of recovery effect. To improve the performance of the algorithm, the traditional Laplace operator is replaced by the reconciliation model, which overcomes the smoothing effect of the Laplace operator on the image and effectively preserves the edge information of the image. In order to take into account the noise suppression and edge protection, an adaptive scheme of the algorithm is proposed according to the characteristics of the suboptimal algorithm, which achieves the recovery of the output image of the optical addressing liquid crystal light valve and effectively improves the image quality.

The algorithms proposed in this article are all implemented under the condition that the point spread function is known. It is necessary to identify the point spread function first and then carry out the recovery.

References

1. Wu Y (2003) Research on image recovery algorithm. University of Electronic Science and Technology
2. Banham MR, Katsaggelos AK (1997) Digital image restoration. IEEE Signal Process Mag 14(3):24–41
3. Zou MY (2001) Deconvolution and signal recovery. National Defense Industry Press, Beijing
4. Chen B (2008) Research on the theory and algorithm of adaptive optical image restoration. University of Information Engineering
5. Zhou YT, Chellappa R, Vaid A et al (1988) Image restoration using a neural network. IEEE Trans Acoust Speech Sign Process 36(7):1141–1151
6. Paik JK, Katsaggelos AK (1990) Image restoration using the Hopfield network with nonzero autoconnection. Acoust Speech Sign Process 4:1909–1912
7. Paik JK, Katsaggelos AK (1992) Image restoration using a modified Hopfield Network. IEEE Trans Image Process 1(1):49–63
8. Paik JK, Katsaggelos AK (1990) Parallel and distributed image restoration using a modified Hopfield network. Parall Arch Image Process 298–307
9. Sun Y (2000a) Hopfield neural network based algorithms for image restoration and reconstruction. I. Algorithms and simulations. Sign Process 2105–2118
10. Sun Y (2000b) Hopfield neural network based algorithms for image restoration and reconstruction. II. Performance analysis. Sign Process 2119–2131

11. Sun Y, Li J, Yu S (1995) Improvement on performance of modified Hopfield neural network for image restoration. Image Process Patt Recogn 688–692
12. Wang L, Qi F, Mo Y (1997) Theoretical and experimental analyses of restoring degraded images based on continuous Hopfield neural networks. Neur Netw 3:1634–1637
13. Han Y, Lenan W (2004) Image restoration using a modified Hopfield neural network of continuous state change. Sign Process 20(5):431–435
14. Perry SW, Guan L (1998) A statistics-based weight assignment in a Hopfield neural network for adaptive image restoration. Neur Netw Proc 2:922–927
15. Yang HC, Wilson R (1995) Adaptive image restoration using a multiresolution Hopfield neural network. Image Process Appl 1995:198–202
16. Ghennam S, Benmahammed K (2001) Adaptive image restoration using Hopfield neural network. Neur Netw Sign Process 569–578
17. Wei Q, Huaidong L, Maria K et al (1998) Adaptive neural network for nuclear medicine image restoration. J VLSI Sign Process 18(3):297–315
18. Wong HS, Guan L (1997) Adaptive regularization in image restoration using a model-based neural network. Opt Eng 36(12):125–136
19. Muneyasu M, Yamamoto K, Hinamoto T (1995) Image restoration using layered neural networks and Hopfield networks. Image Process 33–36
20. Bilal M, Sharif M, Arfan Jaffar M et al (2010) Image restoration using modified Hopfield fuzzy regularization method. Fut Inform Technol 758–763
21. Yu S, Medvedev S, Jaime R (2000) Sensor and method fusion in remote sensing imagery with neural networks. Ant Prop Soc Int Symp 4:1960–1963
22. Perry SW, Guan L (2000) Weight assignment for adaptive image restoration by neural networks. Neur Netw 156–170
23. Liu P, Zhang Y, Mao ZG (2006) Adaptive neural network image restoration algorithm based on symbiotic matrix analysis. J Comp Aid Des Graph 18(8):1205–1211
24. Wu Y, Zhang H (2006) A novel image restoration algorithm using neural network based on variational PDE model. Commun Circ Syst Proc 433–436
25. Zhang HY, Sun Y, Wu YD (2008) The diffusion performance analysis of variational PDE based neural network restoration models. Commun Circ Syst 834–837
26. Wu YD, Sun Y, Zhang HY et al (2007) Variational PDE based image restoration using neural network. Image Process 85–93
27. Wu Y, Zhang H, Sun Y et al (2010) A parallel image restoration algorithm based on Harmonic model using neural networks. Ubi-media Comput 30–33
28. Wu YD, Sun SX (2007) Fast neural network image recovery algorithm based on reconciliation model. Comp Appl Res 24(6):158–160
29. Sun J, Zongben X (2006) An edge preserving regularization model for image restoration based on Hopfield neural network. Lect Notes Comput Sci 3972(1):563–568
30. Zhang H, Yadong W, Peng Q (2005) Image restoration using Hopfield neural network based on total variational model. Lect Notes Comput Sci 3497(1):735–740
31. Lou S, Ding ZL, Yuan F et al (2009) Application of Hopfield neural network and wavelet domain hidden Markov tree model for image recovery. Opt Precis Eng 17(11):2828–2834

Printed in the United States
by Baker & Taylor Publisher Services